Mikhail Moshkov and Beata Zielosko

Combinatorial Machine Learning

## Studies in Computational Intelligence, Volume 360

**Editor-in-Chief**
Prof. Janusz Kacprzyk
Systems Research Institute
Polish Academy of Sciences
ul. Newelska 6
01-447 Warsaw
Poland
*E-mail:* kacprzyk@ibspan.waw.pl

---

Further volumes of this series can be found on our homepage: springer.com

Vol. 340. Heinrich Hussmann, Gerrit Meixner, and Detlef Zuehlke (Eds.)
*Model-Driven Development of Advanced User Interfaces*, 2011
ISBN 978-3-642-14561-2

Vol. 341. Stéphane Doncieux, Nicolas Bredeche, and Jean-Baptiste Mouret(Eds.)
*New Horizons in Evolutionary Robotics*, 2011
ISBN 978-3-642-18271-6

Vol. 342. Federico Montesino Pouzols, Diego R. Lopez, and Angel Barriga Barros
*Mining and Control of Network Traffic by Computational Intelligence*, 2011
ISBN 978-3-642-18083-5

Vol. 343. Kurosh Madani, António Dourado Correia, Agostinho Rosa, and Joaquim Filipe (Eds.)
*Computational Intelligence*, 2011
ISBN 978-3-642-20205-6

Vol. 344. Atilla Elçi, Mamadou Tadiou Koné, and Mehmet A. Orgun (Eds.)
*Semantic Agent Systems*, 2011
ISBN 978-3-642-18307-2

Vol. 345. Shi Yu, Léon-Charles Tranchevent, Bart De Moor, and Yves Moreau
*Kernel-based Data Fusion for Machine Learning*, 2011
ISBN 978-3-642-19405-4

Vol. 346. Weisi Lin, Dacheng Tao, Janusz Kacprzyk, Zhu Li, Ebroul Izquierdo, and Haohong Wang (Eds.)
*Multimedia Analysis, Processing and Communications*, 2011
ISBN 978-3-642-19550-1

Vol. 347. Sven Helmer, Alexandra Poulovassilis, and Fatos Xhafa
*Reasoning in Event-Based Distributed Systems*, 2011
ISBN 978-3-642-19723-9

Vol. 348. Beniamino Murgante, Giuseppe Borruso, and Alessandra Lapucci (Eds.)
*Geocomputation, Sustainability and Environmental Planning*, 2011
ISBN 978-3-642-19732-1

Vol. 349. Vitor R. Carvalho
*Modeling Intention in Email*, 2011
ISBN 978-3-642-19955-4

Vol. 350. Thanasis Daradoumis, Santi Caballé, Angel A. Juan, and Fatos Xhafa (Eds.)
*Technology-Enhanced Systems and Tools for Collaborative Learning Scaffolding*, 2011
ISBN 978-3-642-19813-7

Vol. 351. Ngoc Thanh Nguyen, Bogdan Trawiński, and Jason J. Jung (Eds.)
*New Challenges for Intelligent Information and Database Systems*, 2011
ISBN 978-3-642-19952-3

Vol. 352. Nik Bessis and Fatos Xhafa (Eds.)
*Next Generation Data Technologies for Collective Computational Intelligence*, 2011
ISBN 978-3-642-20343-5

Vol. 353. Igor Aizenberg
*Complex-Valued Neural Networks with Multi-Valued Neurons*, 2011
ISBN 978-3-642-20352-7

Vol. 354. Ljupco Kocarev and Shiguo Lian (Eds.)
*Chaos-Based Cryptography*, 2011
ISBN 978-3-642-20541-5

Vol. 355. Yan Meng and Yaochu Jin (Eds.)
*Bio-Inspired Self-Organizing Robotic Systems*, 2011
ISBN 978-3-642-20759-4

Vol. 356. Slawomir Koziel and Xin-She Yang (Eds.)
*Computational Optimization, Methods and Algorithms*, 2011
ISBN 978-3-642-20858-4

Vol. 357. Nadia Nedjah, Leandro Santos Coelho, Viviana Cocco Mariani, and Luiza de Macedo Mourelle (Eds.)
*Innovative Computing Methods and Their Applications to Engineering Problems*, 2011
ISBN 978-3-642-20957-4

Vol. 358. Norbert Jankowski, Włodzisław Duch, and Krzysztof Grąbczewski (Eds.)
*Meta-Learning in Computational Intelligence*, 2011
ISBN 978-3-642-20979-6

Vol. 359. Xin-She Yang and Slawomir Koziel (Eds.)
*Computational Optimization and Applications in Engineering and Industry*, 2011
ISBN 978-3-642-20985-7

Vol. 360. Mikhail Moshkov and Beata Zielosko
*Combinatorial Machine Learning*, 2011
ISBN 978-3-642-20994-9

Mikhail Moshkov and Beata Zielosko

# Combinatorial Machine Learning

## A Rough Set Approach

**Authors**

Mikhail Moshkov
Mathematical and Computer Sciences and Engineering Division
King Abdullah University of Science and Technology
Thuwal, 23955-6900
Saudi Arabia
E-mail: mikhail.moshkov@kaust.edu.sa

Beata Zielosko
Mathematical and Computer Sciences and Engineering Division
King Abdullah University of Science and Technology
Thuwal, 23955-6900
Saudi Arabia
E-mail: beata.zielosko@kaust.edu.sa

and

Institute of Computer Science
University of Silesia
39, Będzińska St.
Sosnowiec, 41-200
Poland

ISBN 978-3-642-26901-1

ISBN 978-3-642-20995-6 (eBook)

DOI 10.1007/978-3-642-20995-6

Studies in Computational Intelligence ISSN 1860-949X

© 2011 Springer-Verlag Berlin Heidelberg
Softcover reprint of the hardcover 1st edition 2011

This work is subject to copyright. All rights are reserved, whether the whole or part of the material is concerned, specifically the rights of translation, reprinting, reuse of illustrations, recitation, broadcasting, reproduction on microfilm or in any other way, and storage in data banks. Duplication of this publication or parts thereof is permitted only under the provisions of the German Copyright Law of September 9, 1965, in its current version, and permission for use must always be obtained from Springer. Violations are liable to prosecution under the German Copyright Law.

The use of general descriptive names, registered names, trademarks, etc. in this publication does not imply, even in the absence of a specific statement, that such names are exempt from the relevant protective laws and regulations and therefore free for general use.

*Typeset & Cover Design:* Scientific Publishing Services Pvt. Ltd., Chennai, India.

Printed on acid-free paper

9 8 7 6 5 4 3 2 1

springer.com

To our families

# Preface

Decision trees and decision rule systems are widely used in different applications as algorithms for problem solving, as predictors, and as a way for knowledge representation. Reducts play key role in the problem of attribute (feature) selection.

The aims of this book are the consideration of the sets of decision trees, rules and reducts; study of relationships among these objects; design of algorithms for construction of trees, rules and reducts; and deduction of bounds on their complexity. We consider also applications for supervised machine learning, discrete optimization, analysis of acyclic programs, fault diagnosis and pattern recognition.

We study mainly time complexity in the worst case of decision trees and decision rule systems. We consider both decision tables with one-valued decisions and decision tables with many-valued decisions. We study both exact and approximate trees, rules and reducts. We investigate both finite and infinite sets of attributes.

This is a mixture of research monograph and lecture notes. It contains many unpublished results. However, proofs are carefully selected to be understandable. The results considered in this book can be useful for researchers in machine learning, data mining and knowledge discovery, especially for those who are working in rough set theory, test theory and logical analysis of data. The book can be used under the creation of courses for graduate students.

Thuwal, Saudi Arabia
March 2011

Mikhail Moshkov
Beata Zielosko

# Acknowledgements

We are greatly indebted to King Abdullah University of Science and Technology and especially to Professor David Keyes and Professor Brian Moran for various support.

We are grateful to Professor Andrzej Skowron for stimulated discussions and to Czesław Zielosko for the assistance in preparation of figures for the book.

We extend an expression of gratitude to Professor Janusz Kacprzyk, to Dr. Thomas Ditzinger and to the Studies in Computational Intelligence staff at Springer for their support in making this book possible.

# Contents

# Introduction

This book is devoted mainly to the study of decision trees, decision rules and tests (reducts) [8, 70, 71, 90]. These constructions are widely used in supervised machine learning [23] to predict the value of decision attribute for a new object given by values of conditional attributes, in data mining and knowledge discovery to represent knowledge extracted from decision tables (datasets), and in different applications as algorithms for problem solving. In the last case, decision trees should be considered as serial algorithms, but decision rule systems allow parallel implementation.

A test is a subset of conditional attributes which give us the same information about the decision attribute as the whole set of conditional attributes. A reduct is an uncancelable test. Tests and reducts play a special role: their study allow us to choose relevant to our goals sets of conditional attributes (features).

We study decision trees, rules and tests as combinatorial objects: we try to understand the structure of sets of tests (reducts), trees and rules, consider relationships among these objects, design algorithms for construction and optimization of trees, rules and tests, and derive bounds on their complexity.

We concentrate on minimization of the depth of decision trees, length of decision rules and cardinality of tests. These optimization problems are connected mainly with the use of trees and rules as algorithms. They have sense also from the point of view of knowledge representation: decision trees with small depth and short decision rules are more understandable. These optimization problems are associated also with minimum description length principle [72] and, probably, can be useful for supervised machine learning.

The considered subjects are closely connected with machine learning [23, 86]. Since we avoid the consideration of statistical approaches, we hope that Combinatorial Machine Learning is a relevant label for our study. We need to clarify also the subtitle A Rough Set Approach. The three theories are nearest to our investigations: test theory [84, 90, 92], rough set theory [70, 79, 80], and logical analysis of data [6, 7, 17]. However, the rough set theory is more appropriate for this book: only in this theory inconsistent decision tables are

M. Moshkov and B. Zielosko: Combinatorial Machine Learning, SCI 360, pp. 1–3.
springerlink.com © Springer-Verlag Berlin Heidelberg 2011

studied systematically. In such tables there exist rows (objects) with the same values of conditional attributes but different values of the decision attribute. In this book, we consider inconsistent decision tables in the frameworks of decision tables with many-valued decisions.

The monograph contains Introduction, Chap. 1 with main notions and simple examples from different areas of applications, and two parts: Tools and Applications.

The part Tools consists of five chapters (Chaps. 2–6). In Chaps. 2, 3 and 4 we study decision tables with one-valued decisions. We assume that rows of the table are pairwise different, and (for simplicity) we consider only binary conditional attributes. In Chap. 2, we study the structure of sets of decision trees, rules and tests, and relationships among these objects. In Chap. 3, we consider lower and upper bounds on complexity of trees, rules and tests. In Chap. 4, we study both approximate and exact (based on dynamic programming) algorithms for minimization of the depth of trees, length of rules, and cardinality of tests.

In the next two chapters, we continue this line of research: relationships among trees, rules and tests, bounds on complexity and algorithms for construction of these objects. In Chap. 5, we study decision tables with many-valued decisions when each row is labeled not with one value of the decision attribute but with a set of values. Our aim in this case is to find at least one value of the decision attribute. This is a new approach for the rough set theory. Chapter 6 is devoted to the consideration of approximate trees, rules and tests. Their use (instead of exact ones) allows us sometimes to obtain more compact description of knowledge contained in decision tables, and to design more precise classifiers.

The second part Applications contains four chapters. In Chap. 7, we discuss the use of trees, rules and tests in supervised machine learning, including lazy learning algorithms. Chapter 8 is devoted to the study of infinite systems of attributes based on local and global approaches. Local means that we can use in decision trees and decision rule systems only attributes from the problem description. Global approach allows the use of arbitrary attributes from the given infinite system. Tools considered in the first part of the book make possible to understand the behavior in the worst case of the minimum complexity of classifiers based on decision trees and rules, depending on the number of attributes in the problem description.

In Chap. 9, we study decision trees with so-called quasilinear and linear attributes, and applications of obtained results to problems of discrete optimization and analysis of acyclic programs. In particular, we discuss the existence of a decision tree with linear attributes which solves traveling salesman problem with $n \geq 4$ cities and which depth is at most $n^7$. In Chap. 10, we consider two more applications: the diagnosis of constant faults in combinatorial circuits and the recognition of regular language words.

This book is a mixture of research monograph and lecture notes. We tried to systematize tools for the work with exact and approximate decision trees, rules and tests for decision tables both with one-valued and many-valued decisions. To fill various gaps during the systematization we were forced to add a number of unpublished results. However, we selected results and especially proofs carefully to make them understandable for graduate students.

The first course in this direction was taught in Russia in 1984. It covered different topics connected with decision trees and tests for decision tables with one-valued decisions. In 2005 in Poland topics connected with approximate trees and tests as well as decision tables with many-valued decisions were added to a new version of the course. After publishing a series of papers about partial covers, reducts, and decision and association rules [57, 58, 60, 61, 62, 63, 69, 93, 94, 95, 96] including monograph [59], the authors decided to add decision rules to the course. This book is an essential extension of the course Combinatorial Machine Learning in King Abdullah University of Science and Technology (KAUST) in Saudi Arabia.

The results considered in this book can be useful for researchers in machine learning, data mining and knowledge discovery, especially for those who are working in rough set theory, test theory and logical analysis of data. The book can be used for creation of courses for graduate students.

# 1

# Examples from Applications

In this chapter, we discuss briefly main notions: decision trees, rules, complete systems of decision rules, tests and reducts for problems and decision tables.

After that we concentrate on consideration of simple examples from different areas of applications: fault diagnosis, computational geometry, pattern recognition, discrete optimization and analysis of experimental data.

These examples allow us to clarify relationships between problems and corresponding decision tables, and to hint at tools required for analysis of decision tables.

The chapter contains four sections. In Sect. 1.1 main notions connected with problems are discussed. Section 1.2 is devoted to the consideration of main notions connected with decision tables. Section 1.3 contains seven examples, and Sect. 1.4 includes conclusions.

## 1.1 Problems

We begin with simple and important model of a *problem*. Let $A$ be a set (set of inputs or the universe). It is possible that $A$ is an infinite set. Let $f_1, \dots, f_n$ be attributes, each of which is a function from $A$ to $\{0, 1\}$. Each attribute divides the set $A$ into two domains. In the first domain the value of the considered attribute is equal to 0, and in the second domain the value of this attribute is equal to 1 (see Fig. 1.1).

All attributes $f_1, \dots, f_n$ divide the set $A$ into a number of domains in each of which values of attributes are constant. These domains are enumerated such that different domains can have the same number (see Fig. 1.2).

We will consider the following problem: for a given element $a \in A$ it is required to recognize the number of domain to which $a$ belongs. To this end we can use values of attributes from the set $\{f_1, \dots, f_n\}$ on $a$.

More formally, a *problem* is a tuple $(\nu, f_1, \dots, f_n)$ where $\nu$ is a mapping from $\{0, 1\}^n$ to $\mathbb{N}$ (the set of natural numbers) which enumerates the

M. Moshkov and B. Zielosko: Combinatorial Machine Learning, SCI 360, pp. 5–20.
springerlink.com
© Springer-Verlag Berlin Heidelberg 2011

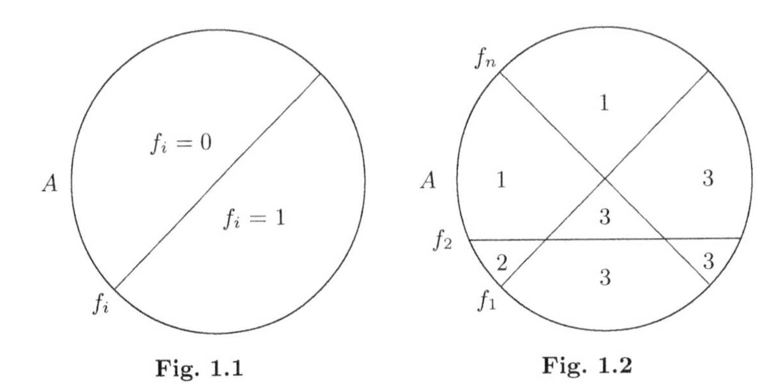


**Fig. 1.1** **Fig. 1.2**

domains. Each domain corresponds to the nonempty set of solutions on $A$ of a set of equations of the kind

$$\{f_1(x) = \delta_1, \ldots, f_n(x) = \delta_n\}$$

where $\delta_1, \ldots, \delta_n \in \{0, 1\}$. The considered problem can be reformulated in the following way: for a given $a \in A$ we should find the number

$$z(a) = \nu(f_1(a), \ldots, f_n(a)) .$$

As algorithms for the considered problem solving we will use decision trees and decision rule systems.

A *decision tree* is a finite directed tree with the root in which each terminal node is labeled with a number (decision), each nonterminal node (such nodes will be called *working* nodes) is labeled with an attribute from the set $\{f_1, \ldots, f_n\}$. Two edges start in each working node. These edges are labeled with 0 and 1 respectively (see Fig. 1.3).

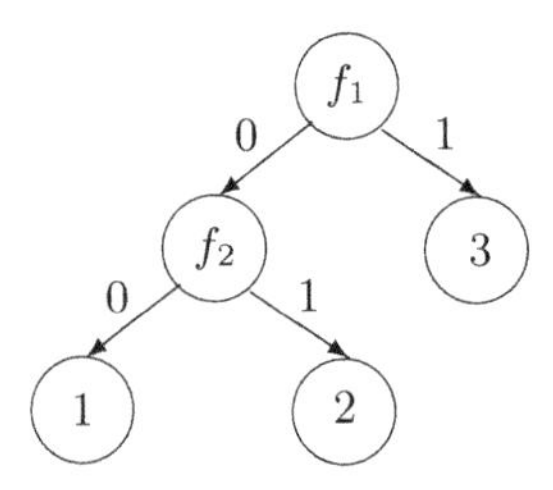


**Fig. 1.3**

Let $\Gamma$ be a decision tree. For a given element $a \in A$ the tree works in the following way: if the root of $\Gamma$ is a terminal node labeled with a number $m$ then $m$ is the result of the tree $\Gamma$ work on the element $a$. Let the root of $\Gamma$

be a working node labeled with an attribute $f_i$. Then we compute the value $f_i(a)$ and pass along the edge labeled with $f_i(a)$, etc.

We will say that $\Gamma$ *solves* the considered problem if for any $a \in A$ the result of $\Gamma$ work coincides with the number of domain to which $a$ belongs.

As time complexity of $\Gamma$ we will consider the *depth* $h(\Gamma)$ of $\Gamma$ which is the maximum length of a path from the root to a terminal node of $\Gamma$. We denote by $h(z)$ the minimum depth of a decision tree which solves the problem $z$.

A *decision rule* $r$ over $z$ is an expression of the kind

$$f_{i_1} = b_1 \wedge \ldots \wedge f_{i_m} = b_m \rightarrow t$$

where $f_{i_1}, \ldots, f_{i_m} \in \{f_1, \ldots, f_n\}$, $b_1, \ldots, b_m \in \{0, 1\}$, and $t \in \mathbb{N}$. The number $m$ is called the *length* of the rule $r$. This rule is called *realizable* for an element $a \in A$ if

$$f_{i_1}(a) = b_1, \ldots, f_{i_m}(a) = b_m .$$

The rule $r$ is called *true* for $z$ if for any $a \in A$ such that $r$ is realizable for $a$, the equality $z(a) = t$ holds.

A *decision rule system* $S$ over $z$ is a nonempty finite set of rules over $z$. A system $S$ is called a *complete* decision rule system for $z$ if each rule from $S$ is true for $z$, and for every $a \in A$ there exists a rule from $S$ which is realizable for $a$. We can use a complete decision rule system $S$ to solve the problem $z$. For a given $a \in A$ we should find a rule $r \in S$ which is realizable for $a$. Then the number from the right-hand side of $r$ is equal to $z(a)$.

We denote by $L(S)$ the maximum length of a rule from $S$, and by $L(z)$ we denote the minimum value of $L(S)$ among all complete decision rule systems $S$ for $z$. The value $L(S)$ can be interpreted as time complexity in the worst case of the problem $z$ solving by $S$ if we have their own processor for each rule from $S$.

Except of decision trees and decision rule systems we will consider tests and reducts. A *test* for the problem $z = (\nu, f_1, \ldots, f_n)$ is a subset $\{f_{i_1}, \ldots, f_{i_m}\}$ of the set $\{f_1, \ldots, f_n\}$ such that there exists a mapping $\mu : \{0, 1\}^m \rightarrow \mathbb{N}$ for which

$$\nu(f_1(a), \ldots, f_n(a)) = \mu(f_{i_1}(a), \ldots, f_{i_m}(a))$$

for any $a \in A$. In the other words, test is a subset of the set of attributes $\{f_1, \ldots, f_n\}$ such that values of the considered attributes on any element $a \in A$ are enough for the problem $z$ solving on the element $a$. A *reduct* is a test such that each proper subset of this test is not a test for the problem. It is clear that each test has a reduct as a subset. We denote by $R(z)$ the minimum cardinality of a reduct for the problem $z$.

## 1.2 Decision Tables

We associate a *decision table* $T = T(z)$ with the considered problem (see Fig. 1.4).

| | $f_1$ | ... | $f_n$ | |
|---|---|---|---|---|
| $T=$ | $\delta_1$ | ... | $\delta_n$ | $\nu(\delta_1, \ldots, \delta_n)$ |

**Fig. 1.4**

This table is a rectangular table with $n$ columns corresponding to attributes $f_1, \ldots, f_n$. A tuple $(\delta_1, \ldots, \delta_n) \in \{0,1\}^n$ is a row of $T$ if and only if the system of equations

$$\{f_1(x) = \delta_1, \ldots, f_n(x) = \delta_n\}$$

is compatible on the set $A$ (has a solution on the set $A$). This row is labeled with the number $\nu(\delta_1, \ldots, \delta_n)$.

We can correspond a *game* of two players to the table $T$. The first player chooses a row of the table $T$ and the second one should recognize the number (decision) attached to this row. To this end the second player can choose columns of $T$ and ask the first player what is at the intersection of these columns and the considered row. The *strategies of the second player* can be represented in the form of decision trees or decision rule systems.

It is not difficult to show that the set of strategies of the second player represented in the form of decision trees coincides with the set of decision trees with attributes from $\{f_1, \ldots, f_n\}$ solving the problem $z = (\nu, f_1, \ldots, f_n)$. We denote by $h(T)$ the minimum depth of decision tree for the table $T = T(z)$ which is a strategy of the second player. It is clear that $h(z) = h(T(z))$.

We can formulate the notion of decision rule over $T$, the notion of decision rule realizable for a row of $T$, and the notion of decision rule true for $T$ in a natural way. We will say that a system $S$ of decision rules over $T$ is a *complete* decision rule system for $T$ if each rule from $S$ is true for $T$, and for every row of $T$ there exists a rule from $S$ which is realizable for this row.

A complete system of rules $S$ can be used by the second player to find the decision attached to the row chosen by the first player. If the second player can work with rules in parallel, the value $L(S)$—the maximum length of a rule from $S$—can be interpreted as time complexity in the worst case of corresponding strategy of the second player. We denote by $L(T)$ the minimum value of $L(S)$ among all complete decision rule systems $S$ for $T$. One can show that a decision rule system $S$ over $z$ is complete for $z$ if and only if $S$ is complete for $T = T(z)$. So $L(z) = L(T(z))$.

We can formulate the notion of test for the table $T$: a set $\{f_{i_1}, \ldots, f_{i_m}\}$ of columns of the table $T$ is a test for the table $T$ if each two rows of $T$ with different decisions are different on at least one column from the set $\{f_{i_1}, \ldots, f_{i_m}\}$. A reduct for the table $T$ is a test for which each proper subset is not a test. We denote by $R(T)$ the minimum cardinality of a reduct for the table $T$.

One can show that a subset of attributes $\{f_{i_1}, \ldots, f_{i_m}\}$ is a test for the problem $z$ if and only if the set of columns $\{f_{i_1}, \ldots, f_{i_m}\}$ is a test for the table $T = T(z)$. It is clear that $R(z) = R(T(z))$.

So instead of the problem $z$ we can study the decision table $T(z)$.

## 1.3 Examples

There are two sources of problems and corresponding decision tables: classes of exactly formulated problems and experimental data. We begin with very simple example about three inverted cups and a small ball under one of these cups. Later, we consider examples of exactly formulated problems from the following areas:

- Diagnosis of faults in combinatorial circuits,
- Computational geometry,
- Pattern recognition,
- Discrete optimization.

The last example is about data table with experimental data.

### *1.3.1 Three Cups and Small Ball*

Let we have three inverted cups on the table and a small ball under one of these cups (see Fig. 1.5).

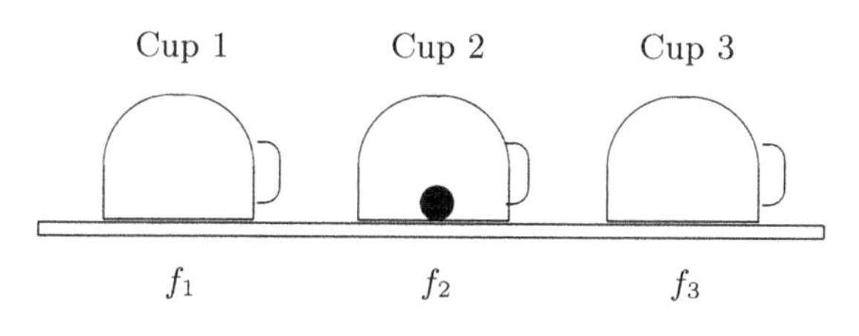


**Fig. 1.5**

We should find a number of cup under which the ball lies. To this end we will use attributes $f_i$, $i = 1, 2, 3$. We are lifting the $i$-th cup. If the ball lies under this cup then the value of $f_i$ is equal to 1. Otherwise, the value of $f_i$ is equal to 0. These attributes are defined on the set $A = \{a_1, a_2, a_3\}$ where $a_i$ is the location of the ball under the $i$-th cup, $i = 1, 2, 3$.

We can represent this problem in the following form: $z = (\nu, f_1, f_2, f_3)$ where $\nu(1,0,0) = 1$, $\nu(0,1,0) = 2$, $\nu(0,0,1) = 3$, and $\nu(\delta_1, \delta_2, \delta_3) = 4$ for any tuple $(\delta_1, \delta_2, \delta_3) \in \{0,1\}^3 \setminus \{(1,0,0), (0,1,0), (0,0,1)\}$. The decision table $T = T(z)$ is represented in Fig. 1.6.

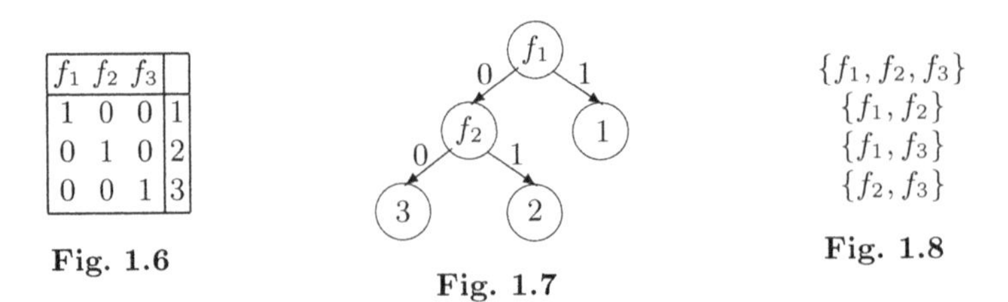

| $f_1$ | $f_2$ | $f_3$ | |
|---|---|---|---|
| 1 | 0 | 0 | 1 |
| 0 | 1 | 0 | 2 |
| 0 | 0 | 1 | 3 |

**Fig. 1.6**



**Fig. 1.7**



**Fig. 1.8**

A decision tree solving this problem is represented in Fig. 1.7, and in Fig. 1.8 all tests for this problem are represented. It is clear that $R(T) = 2$ and $h(T) \leq 2$.

Let us assume that $h(T) = 1$. Then there exists a decision tree which solves $z$ and has a form represented in Fig. 1.9, but it is impossible since this tree has only two terminal nodes, and the considered problem has three different solutions. So $h(z) = h(T) = 2$.

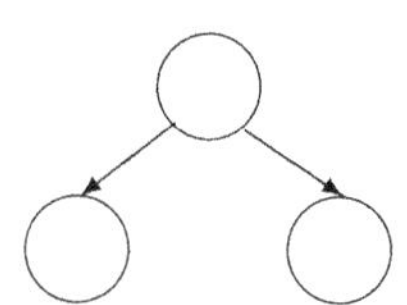

**Fig. 1.9**

One can show that

$$\{f_1 = 1 \to 1, f_2 = 1 \to 2, f_3 = 1 \to 3\}$$

is a complete decision rule system for $T$, and for $i = 1, 2, 3$, the $i$-th rule is the shortest rule which is true for $T$ and realizable for the $i$-th row of $T$. Therefore $L(T) = 1$ and $L(z) = 1$.

### *1.3.2 Diagnosis of One-Gate Circuit*

Let we have a circuit $S$ represented in Fig. 1.10. Each input of the gate $\wedge$ can work correctly or can have constant fault from the set $\{0, 1\}$. For example, the fault 0 on the input $x$ means that independently of the value incoming to the input $x$, this input transmits 0 to the gate $\wedge$.

Each fault of the circuit $S$ can be represented by a tuple from the set $\{0, 1, c\}^2$. For example, the tuple $(c, 1)$ means that the input $x$ works correctly, but $y$ has constant fault 1 and transmits 1.

The circuit $S$ with fault $(c, c)$ (really without faults) realizes the function $x \wedge y$; with fault $(c, 1)$ realizes $x$; with fault $(1, c)$ realizes $y$, with fault $(1, 1)$ realizes 1; and with faults $(c, 0)$, $(0, c)$, $(1, 0)$, $(0, 1)$ and $(0, 0)$ realizes the

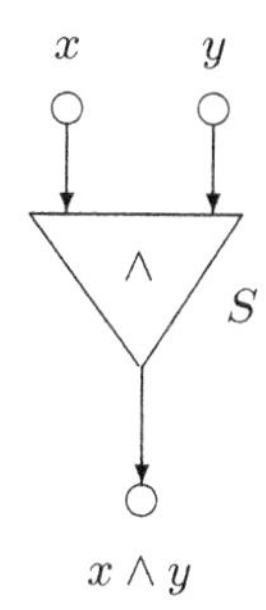


**Fig. 1.10**

function 0. So, if we can only observe the output of $S$ on inputs of which a tuple from $\{0,1\}^2$ is given, then we can not recognize exactly the fault, but we can only recognize the function which the circuit with the fault realizes. The problem of recognition of the function realizing by the circuit $S$ with fault from $\{0,1,c\}^2$ will be called the problem of diagnosis of $S$.

For this problem solving, we will use attributes from the set $\{0,1\}^2$. We give a tuple $(a,b)$ from the set $\{0,1\}^2$ on inputs of $S$ and observe the value on the output of $S$, which is the value of the considered attribute that will be denoted by $f_{ab}$. For the problem of diagnosis, in the capacity of the set $A$ (the universe) we can take the set of circuits $S$ with arbitrary faults from $\{0,1,c\}^2$.

The decision table for the considered problem is represented in Fig. 1.11.

| $f_{00}$ | $f_{01}$ | $f_{10}$ | $f_{11}$ | |
|---|---|---|---|---|
| 0 | 0 | 0 | 1 | $x \wedge y$ |
| 0 | 0 | 1 | 1 | $x$ |
| 0 | 1 | 0 | 1 | $y$ |
| 1 | 1 | 1 | 1 | 1 |
| 0 | 0 | 0 | 0 | 0 |

**Fig. 1.11**

The first and the second rows have different decisions and are different only in the third column. Therefore the attribute $f_{10}$ belongs to each test. The first and the third rows are different only in the second column. Therefore $f_{01}$ belongs to each test. The first and the last rows are different only in the last column. Therefore $f_{11}$ belongs to each test. One can show that $\{f_{01}, f_{10}, f_{11}\}$ is a test. Therefore the considered table has only two tests $\{f_{01}, f_{10}, f_{11}\}$ and $\{f_{00}, f_{01}, f_{10}, f_{11}\}$. Among them only the first test is a reduct. Hence $R(T) = 3$.

The tree depicted in Fig. 1.12 solves the problem of diagnosis of the circuit $S$. Therefore $h(T) \leq 3$.

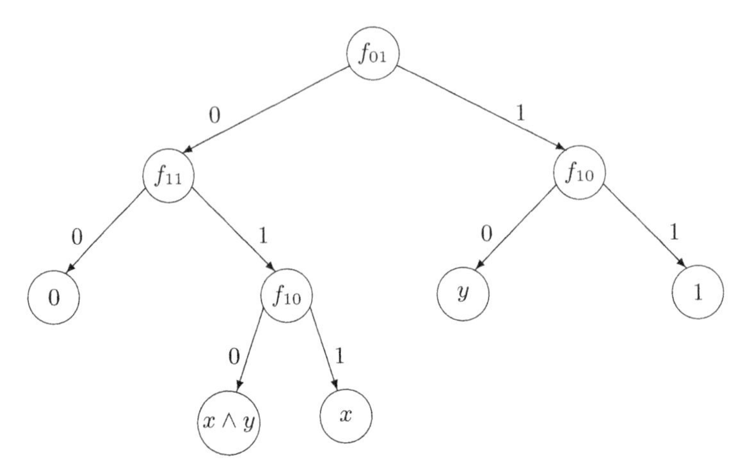


**Fig. 1.12**

Let us assume that $h(T) < 3$. Then there exists a decision tree of the kind depicted in Fig. 1.13, which solves the problem of diagnosis. But this is impossible since there are 5 different decisions and only 4 terminal nodes. So, $h(T) = 3$.

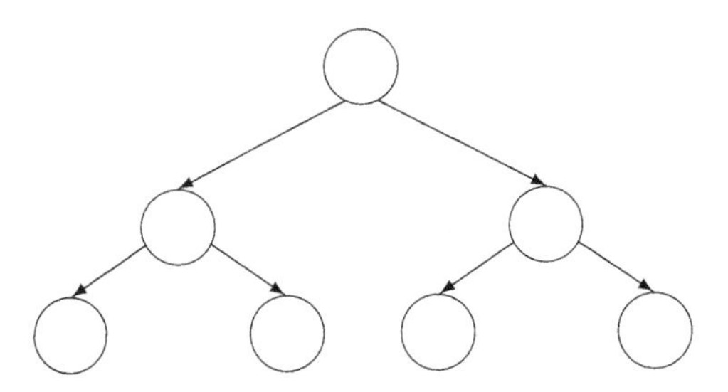

**Fig. 1.13**

One can show that

$$\{f_{01} = 0 \wedge f_{10} = 0 \wedge f_{11} = 1 \to x \wedge y, f_{10} = 1 \wedge f_{00} = 0 \to x,$$
$$f_{01} = 1 \wedge f_{00} = 0 \to y, f_{00} = 1 \to 1, f_{11} = 0 \to 0\}$$

is a complete decision rule system for $T$, and for $i = 1, 2, 3, 4, 5$, the $i$-th rule is the shortest rule which is true for $T$ and realizable for the $i$-th row of $T$. Therefore $L(T) = 3$. It was an example of *fault diagnosis* problem.

### 1.3.3 Problem of Three Post-Offices

Let three post-offices $P_1, P_2$ and $P_3$ exist (see Fig. 1.14). Let new client appear. Then this client will be served by nearest post-office (for simplicity we will assume that the distances between client and post-offices are pairwise distinct).

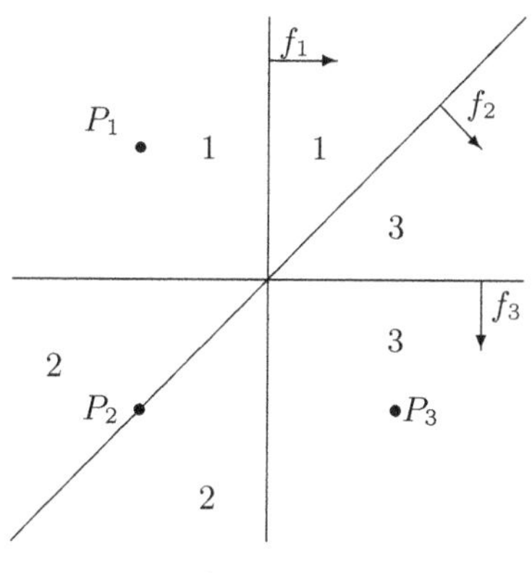


**Fig. 1.14**

Let we have two points $B_1$ and $B_2$. We join these points by segment (of straight line) and draw the perpendicular through the center of this segment (see Fig. 1.15). All points which lie from the left of this perpendicular are

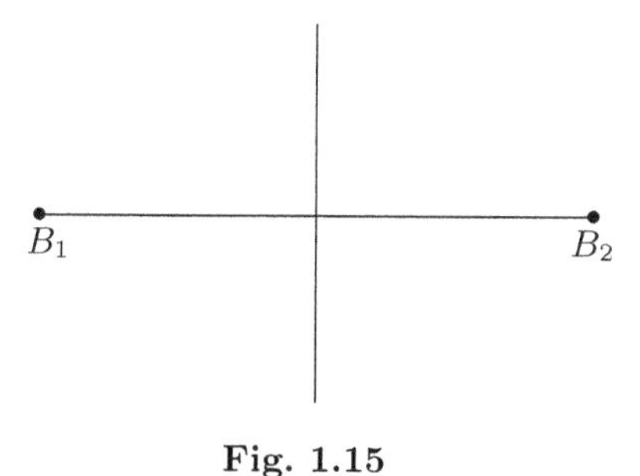


**Fig. 1.15**

nearer to $B_1$, and all points which lie from the right of the perpendicular are nearer to the point $B_2$. This reasoning allows us to construct attributes for the problem of three post-offices.

We joint all pairs of post-offices $P_1, P_2, P_3$ by segments (these segments are invisible in Fig. 1.14) and draw perpendiculars through centers of these segments (note that new client does not belong to these perpendiculars). These perpendiculars (lines) correspond to three attributes $f_1, f_2, f_3$. Each such attribute takes value 0 from the left of the considered line, and takes value 1 from the right of the considered line (arrow points to the right). These three straight lines divide the plane into six regions. We mark each region by the number of post-office which is nearest to points of this region (see Fig. 1.14).

For the considered problem, the set $A$ (the universe) coincides with plane with the exception of these three lines (perpendiculars).

Now we can construct the decision table $T$ corresponding to this problem (see Fig. 1.16).

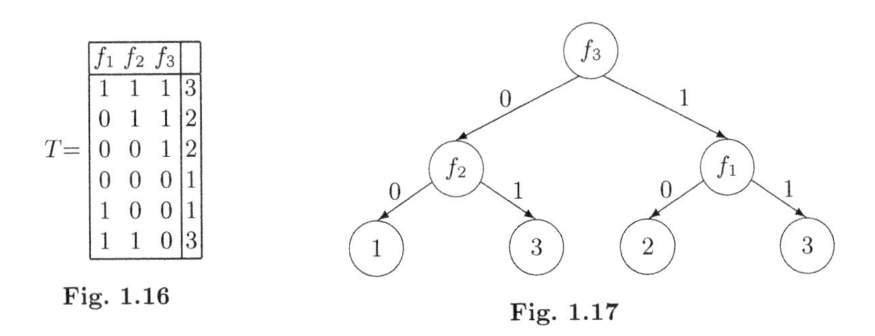

$T=$

| $f_1$ | $f_2$ | $f_3$ | |
|---|---|---|---|
| 1 | 1 | 1 | 3 |
| 0 | 1 | 1 | 2 |
| 0 | 0 | 1 | 2 |
| 0 | 0 | 0 | 1 |
| 1 | 0 | 0 | 1 |
| 1 | 1 | 0 | 3 |



**Fig. 1.16**

**Fig. 1.17**

The first and the second rows of this table have different decisions and are different only in the first column. The fifth and the last rows are different only in the second column and have different decisions. The third and the fourth rows are different only in the third column and have different decisions. So each column of this table belongs to each test. Therefore this table has unique test $\{f_1, f_2, f_3\}$ and $R(T) = 3$.

The decision tree depicted in Fig. 1.17 solves the problem of three post-offices. It is clear that using attributes $f_1, f_2, f_3$ it is impossible to construct a decision tree which depth is equal to 1, and which solves the considered problem. So $h(T) = 2$.

One can show that

$$\{f_1 = 1 \wedge f_2 = 1 \rightarrow 3, f_1 = 0 \wedge f_2 = 1 \rightarrow 2, f_1 = 0 \wedge f_3 = 1 \rightarrow 2,$$
$$f_2 = 0 \wedge f_3 = 0 \rightarrow 1, f_2 = 0 \wedge f_1 = 1 \rightarrow 1, f_1 = 1 \wedge f_2 = 1 \rightarrow 3\}$$

is a complete decision rule system for $T$, and for $i = 1, 2, 3, 4, 5, 6$, the $i$-th rule is the shortest rule which is true for $T$ and realizable for the $i$-th row of $T$. Therefore $L(T) = 2$.

The considered problem is an example of problems studied in *computational geometry*. Note that if we take the plane in the capacity of the universe we will obtain a decision table with many-valued decisions.

### 1.3.4 Recognition of Digits

In Russia, postal address includes six-digit index. On an envelope each digit is drawn on a special matrix (see Figs. 1.18 and 1.19).

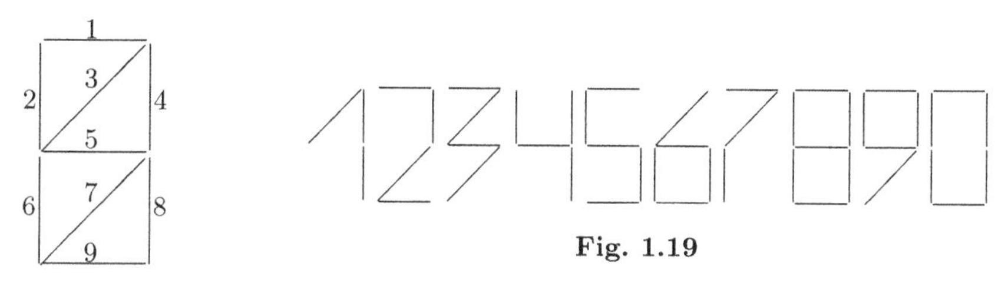


Fig. 1.18

Fig. 1.19

We assume that in the post-office for each element of the matrix there exists a sensor which value is equal to 1 if the considered element is painted and 0 otherwise. So, we have nine two-valued attributes $f_1, ..., f_9$ corresponding to these sensors.

Our aim is to find the minimum number of sensors which are sufficient for recognition of digits. To this end we can construct the decision table, corresponding to the considered problem (see Fig. 1.20). The set $\{f_4, f_5, f_6, f_8\}$

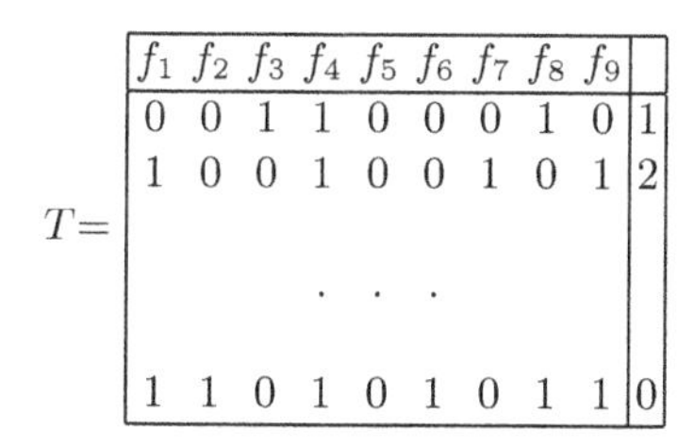

$T=$

| $f_1$ | $f_2$ | $f_3$ | $f_4$ | $f_5$ | $f_6$ | $f_7$ | $f_8$ | $f_9$ | |
|---|---|---|---|---|---|---|---|---|---|
| 0 | 0 | 1 | 1 | 0 | 0 | 0 | 1 | 0 | 1 |
| 1 | 0 | 0 | 1 | 0 | 0 | 1 | 0 | 1 | 2 |
| | | | | . . . | | | | | |
| 1 | 1 | 0 | 1 | 0 | 1 | 0 | 1 | 1 | 0 |

Fig. 1.20

(see Fig. 1.21) is a test for the table $T$. Really, Fig. 1.22 shows that all rows of $T$ are pairwise different at the intersection with columns $f_4, f_5, f_6, f_8$. To simplify the procedure of checking we attached to each digit the number of painted elements with indices from the set $\{4, 5, 6, 8\}$.

Therefore $R(T) \leq 4$. It is clear that we can not recognize 10 objects using only three two-valued attributes. Therefore $R(T) = 4$. It is clear that each decision tree which uses attributes from the set $\{f_1, ..., f_9\}$ and which depth is at most three has at most eight terminal nodes. Therefore $h(T) \geq 4$. The decision tree depicted in Fig. 1.23 solves the considered problem, and the depth of this tree is equal to four. Hence, $h(T) = 4$. It was an example of *pattern recognition* problem.

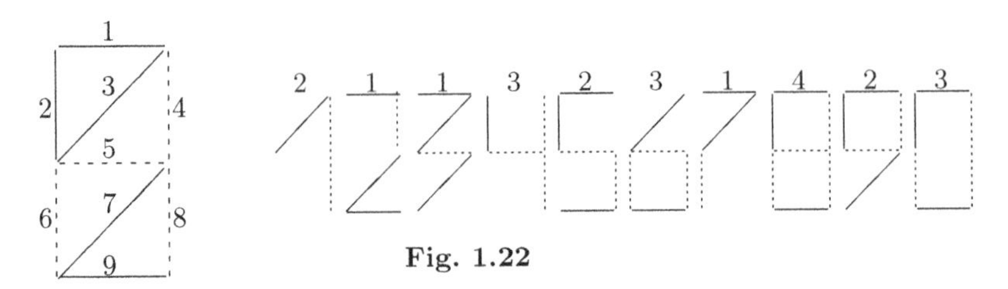


Fig. 1.21

Fig. 1.22

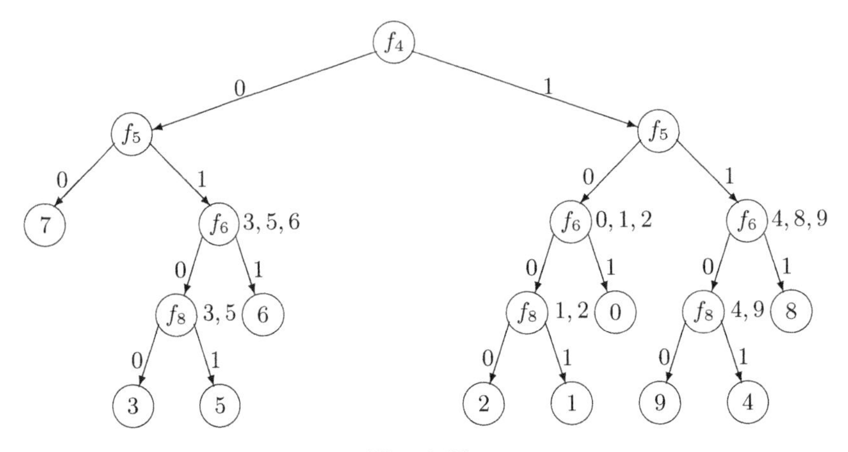


Fig. 1.23

## 1.3.5 Traveling Salesman Problem with Four Cities

Let we have complete unordered graph with four nodes in which each edge is marked by a real number—the length of this edge (see Fig. 1.24).

A Hamiltonian circuit is a closed path which passes through each node exactly one time. We should find a Hamiltonian circuit which has minimum length. There are three Hamiltonian circuits:

$H_1$: 12341 or, which is the same, 14321,

$H_2$: 12431 or 13421,

$H_3$: 13241 or 14231.

For $i = 1, 2, 3$, we denote by $L_i$ the length of $H_i$.

Then

$$L_1 = x_{12} + x_{23} + x_{34} + x_{14} = \overset{\alpha}{(x_{12} + x_{34})} + \overset{\beta}{(x_{23} + x_{14})},$$

$$L_2 = x_{12} + x_{24} + x_{34} + x_{13} = \overset{\alpha}{(x_{12} + x_{34})} + \overset{\gamma}{(x_{24} + x_{13})},$$

$$L_3 = x_{13} + x_{23} + x_{24} + x_{14} = \overset{\gamma}{(x_{24} + x_{13})} + \overset{\beta}{(x_{23} + x_{14})}.$$

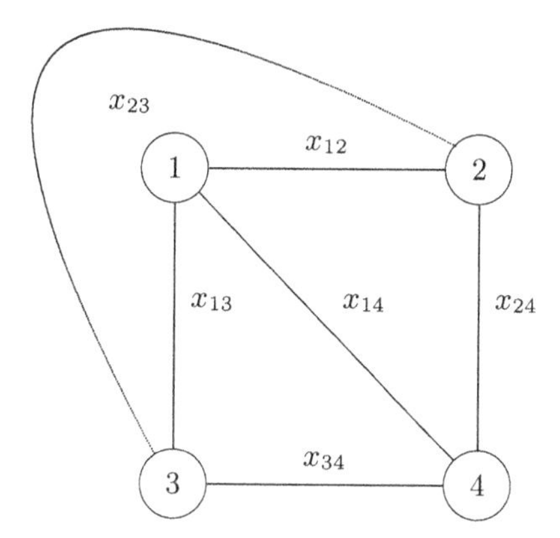


**Fig. 1.24**

For simplicity, we will assume that $L_1, L_2$ and $L_3$ are pairwise different numbers. So, as universe we will consider the set of points of the space $\mathbb{R}^6$ which do not lie on hyperplanes defined by equations $L_1 = L_2$, $L_1 = L_3$, $L_2 = L_3$.

In the capacity of attributes we will use three functions $f_1 = \text{sign}(L_1 - L_2)$, $f_2 = \text{sign}(L_1 - L_3)$, and $f_3 = \text{sign}(L_2 - L_3)$ where $\text{sign}(x) = -1$ if $x < 0$, $\text{sign}(x) = 0$ if $x = 0$, and $\text{sign}(x) = +1$ if $x > 0$. Instead of $+1$ and $-1$ we will write sometimes $+$ and $-$.

Values $L_1$, $L_2$ and $L_3$ are linearly ordered. Let us show that any order is possible. It is clear that values of $\alpha$, $\beta$ and $\gamma$ can be chosen independently.

We can construct corresponding decision table (see Fig. 1.25).

| | $f_1$ | $f_2$ | $f_3$ | |
|---|---|---|---|---|
| If $\alpha < \beta < \gamma$ then $L_1 < L_2 < L_3$ | $-$ | $-$ | $-$ | 1 |
| If $\alpha < \gamma < \beta$ then $L_2 < L_1 < L_3$ | $+$ | $-$ | $-$ | 2 |
| If $\beta < \alpha < \gamma$ then $L_1 < L_3 < L_2$ | $-$ | $-$ | $+$ | 1 |
| If $\beta < \gamma < \alpha$ then $L_3 < L_1 < L_2$ | $-$ | $+$ | $+$ | 3 |
| If $\gamma < \alpha < \beta$ then $L_2 < L_3 < L_1$ | $+$ | $+$ | $-$ | 2 |
| If $\gamma < \beta < \alpha$ then $L_3 < L_2 < L_1$ | $+$ | $+$ | $+$ | 3 |

$= T$

**Fig. 1.25**

We see that the first and second rows have different decisions and are different only in the first column. The third and the fourth rows have different decisions and are different only in the second column. The fifth and the sixth rows have different decisions and are different only in the third column. Therefore $R(T) = 3$. It is clear that $h(T) \geq 2$. A decision tree, represented in Fig. 1.26 solves the considered problem. The depth of this tree is equal to 2. Hence $h(T) = 2$.

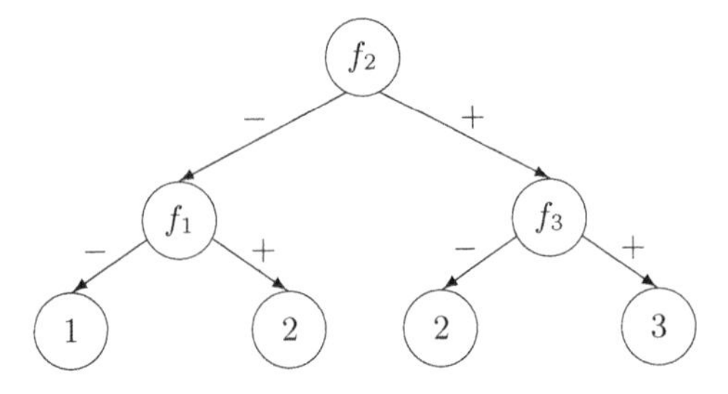


**Fig. 1.26**

One can show that

$$\{f_1 = -1 \wedge f_2 = -1 \rightarrow 1, f_1 = +1 \wedge f_2 = -1 \rightarrow 2,$$
$$f_1 = -1 \wedge f_2 = -1 \rightarrow 1, f_2 = +1 \wedge f_1 = -1 \rightarrow 3,$$
$$f_1 = +1 \wedge f_3 = -1 \rightarrow 2, f_2 = +1 \wedge f_3 = +1 \rightarrow 3\}$$

is a complete decision rule system for $T$, and for $i = 1, 2, 3, 4, 5, 6$, the $i$-th rule is the shortest rule which is true for $T$ and realizable for the $i$-th row of $T$. Therefore $L(T) = 2$. Note that one rule can cover more than one row of decision table (see the first rule, and the first and the third rows).

It was an example of *discrete optimization* problem.

If we consider also points which lie on the mentioned three hyperplanes then we will obtain a decision table with many-valued decisions.

## *1.3.6 Traveling Salesman Problem with $n \geq 4$ Cities*

Until now we have considered so-called local approach to the investigation of decision trees where only attributes from problem description can be used in decision trees and rules. Of course, it is possible to consider global approach too, when we can use arbitrary attributes from the information system in decision trees. Global approach is essentially more complicated than the local one, but in the frameworks of the global approach we sometimes can construct more simple decision trees. Let us consider an example.

Let $G_n$ be the complete unordered graph with $n$ nodes. This graph has $n(n-1)/2$ edges which are marked by real numbers, and $(n-1)!/2$ Hamiltonian circuits. We should find a Hamiltonian circuit with minimum length. This is a problem in the space $\mathbb{R}^{n(n-1)/2}$. What will be if we use for this problem solving arbitrary attributes of the following kind. Let $C$ be an arbitrary hyperplane in $\mathbb{R}^{n(n-1)/2}$. This hyperplane divides the space into two open halfspaces and the hyperplane. The considered attribute takes value $-1$ in one halfspace, value $+1$ in the other halfspace, and the value 0 in the hyperplane.

One can prove that there exists a decision tree using these attributes which solves the considered problem and which depth is at most $n^7$.

One can prove also that for the considered problem there exists a complete decision rule system using these attributes in which the length of each rule is at most $n(n-1)/2+1$.

### 1.3.7 Data Table with Experimental Data

As it was said earlier, there are two sources of decision tables: exactly formulated problems and experimental or statistical data. Now we consider an example of *experimental data.*

Let we have data table (see Fig. 1.27) filled by some experimental data.

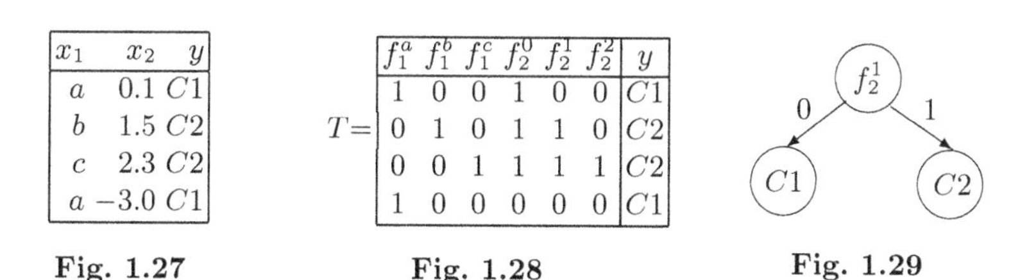

| $x_1$ | $x_2$ | $y$ |
|---|---|---|
| $a$ | 0.1 | $C1$ |
| $b$ | 1.5 | $C2$ |
| $c$ | 2.3 | $C2$ |
| $a$ | −3.0 | $C1$ |

Fig. 1.27

$T=$

| $f_1^a$ | $f_1^b$ | $f_1^c$ | $f_2^0$ | $f_2^1$ | $f_2^2$ | $y$ |
|---|---|---|---|---|---|---|
| 1 | 0 | 0 | 1 | 0 | 0 | $C1$ |
| 0 | 1 | 0 | 1 | 1 | 0 | $C2$ |
| 0 | 0 | 1 | 1 | 1 | 1 | $C2$ |
| 1 | 0 | 0 | 0 | 0 | 0 | $C1$ |

Fig. 1.28



Fig. 1.29

Here $x_1$ and $y$ are discrete variables which take values from some unordered sets, and $x_2$ is a continuous variable. We should predict value of $y$ using variables $x_1$ and $x_2$. We will not use values of variables $x_1$ and $x_2$ directly, but we will use some attributes depending on these variables. We consider attributes which are used in well known system CART [8].

For discrete variable $x_1$, we can take a subset $B$ of the set $\{a, b, c\}$. Then the considered attribute has value 0 if $x_1 \notin B$, and has value 1 if $x_1 \in B$.

Let $f_1^a$ be the attribute corresponding to $B = \{a\}$, $f_1^b$ be the attribute corresponding to $B = \{b\}$, and $f_1^c$ be the attribute corresponding to $B = \{c\}$.

For continuous variable $x_2$, we consider linear ordering of values of this variable $-3.0 < 0.1 < 1.5 < 2.3$ and take some real numbers which lie between neighboring pairs of values, for example, 0, 1 and 2. Let $\alpha$ be such a number. Then the considered attribute takes value 0 if $x_2 < \alpha$, and takes value 1 if $x_2 \geq \alpha$.

Let $f_2^0$, $f_2^1$, and $f_2^2$ be attributes corresponding to numbers 0, 1 and 2 respectively. The decision table for the considered attributes based on variables $x_1$, and $x_2$ is depicted in Fig. 1.28.

We see that $\{f_2^1\}$ is a reduct for this table. Therefore $R(T) = 1$. It is clear that $h(T) = 1$ (see decision tree depicted in Fig. 1.29).

One can show that

$$\{f_1^a = 1 \to C1, f_1^a = 0 \to C2, f_1^a = 0 \to C2, f_1^a = 1 \to C1\}$$

is a complete decision rule system for $T$, and for $i = 1, 2, 3, 4$, the $i$-th rule is the shortest rule which is true for $T$ and realizable for the $i$-th row of $T$.

Therefore $L(T) = 1$. We have here one more example of the situation when one rule covers more than one row of decision table.

## 1.4 Conclusions

The chapter is devoted to brief consideration of main notions and discussion of examples from various areas of applications: fault diagnosis, computational geometry, pattern recognition, discrete optimization, and analysis of experimental data.

The main conclusion is that the study of miscellaneous problems can be reduced to the study of in some sense similar objects—decision tables.

Note that in two examples (problem of three post-offices and traveling salesman problem) we did not consider some inputs. If we eliminate these restrictions we will obtain decision tables with many-valued decisions.

Next five chapters are devoted to the creation of tools for study of decision tables including tables with many-valued decisions.

In Chaps. 2, 3 and 4, we study decision tables with one-valued decisions. In Chap. 2, we consider sets of decision trees, rules and reducts, and relationships among these objects. Chapter 3 deals with bounds on complexity and Chap. 4—with algorithms for construction of trees, rules and reducts.

Chapters 5 and 6 contain two extensions of this study. In Chap. 5, we consider decision tables with many-valued decisions, and in Chap. 6—approximate decision trees, rules and reducts.

# Part I

## Tools

# 2

# Sets of Tests, Decision Rules and Trees

As we have seen, decision tables arise in different applications. So, we study decision tables as an independent mathematical object. We begin our consideration from decision tables with one-valued decisions. For simplicity, we deal mainly with decision tables containing only binary conditional attributes.

This chapter is devoted to the study of the sets of tests (reducts), decision rules and trees. For tests and rules we concentrate on consideration of so-called characteristic functions—monotone Boolean functions that represent sets of tests and rules. We can't describe the set of decision trees in the same way, but we can compare efficiently sets of decision trees for two decision tables with the same attributes. We study also relationships among trees, rules and tests.

The chapter consists of four sections. In Sect. 2.1, main notions are discussed. In Sect. 2.2, the sets of tests, decision rules and trees are studied. In Sect. 2.3, relationships among trees, rules and tests are considered. Section 2.4 contains conclusions.

## 2.1 Decision Tables, Trees, Rules and Tests

A *decision table* is a rectangular table which elements belong to the set $\{0, 1\}$ (see Fig. 2.1). Columns of this table are labeled with attributes $f_1, ..., f_n$. Rows of the table are pairwise different, and each row is labeled with a natural number (a decision). This is a table with one-valued decisions.

$T =$

| $f_1$ | $\ldots$ | $f_n$ | |
|---|---|---|---|
| $\delta_1$ | $\ldots$ | $\delta_n$ | $d$ |

**Fig. 2.1**

M. Moshkov and B. Zielosko: Combinatorial Machine Learning, SCI 360, pp. 23–36.
springerlink.com © Springer-Verlag Berlin Heidelberg 2011

We will associate *a game* of two players with this table. The first player chooses a row of the table and the second player must recognize a decision corresponding to this row. To this end he can choose columns (attributes) and ask the first player what is at the intersection of the considered row and these columns.

A *decision tree over* $T$ is a finite tree with root in which each terminal node is labeled with a decision (a natural number), each nonterminal node (such nodes will be called *working*) is labeled with an attribute from the set $\{f_1, \ldots, f_n\}$. Two edges start in each working node. These edges are labeled with 0 and 1 respectively.

Let $\Gamma$ be a decision tree over $T$. For a given row $r$ of $T$ this tree works in the following way. We begin the work in the root of $\Gamma$. If the considered node is terminal then the result of $\Gamma$ work is the number attached to this node. Let the considered node be working node which is labeled with an attribute $f_i$. If the value of $f_i$ in the considered row is 0 then we pass along the edge which is labeled with 0. Otherwise, we pass along the edge which is labeled with 1, etc.

We will say that $\Gamma$ is a *decision tree for* $T$ if for any row of $T$ the work of $\Gamma$ finishes in a terminal node, which is labeled with the decision corresponding to the considered row.

We denote by $h(\Gamma)$ the *depth* of $\Gamma$ which is the maximum length of a path from the root to a terminal node. We denote by $h(T)$ the minimum depth of a decision tree for the table $T$.

A *decision rule over* $T$ is an expression of the kind

$$f_{i_1} = b_1 \wedge \ldots \wedge f_{i_m} = b_m \rightarrow t$$

where $f_{i_1}, \ldots, f_{i_m} \in \{f_1, \ldots, f_n\}$, $b_1, \ldots, b_m \in \{0, 1\}$, and $t \in \mathbb{N}$. The number $m$ is called the *length* of the rule. This rule is called *realizable* for a row $r = (\delta_1, \ldots, \delta_n)$ if

$$\delta_{i_1} = b_1, \ldots, \delta_{i_m} = b_m \ .$$

The rule is called *true* for $T$ if for any row $r$ of $T$, such that the rule is realizable for row $r$, the row $r$ is labeled with the decision $t$. We denote by $L(T, r)$ the minimum length of a rule over $T$ which is true for $T$ and realizable for $r$. We will say that the considered rule is a *rule for* $T$ *and* $r$ if this rule is true for $T$ and realizable for $r$.

A *decision rule system* $S$ *over* $T$ is a nonempty finite set of rules over $T$. A system $S$ is called a *complete decision rule system for* $T$ if each rule from $S$ is true for $T$, and for every row of $T$ there exists a rule from $S$ which is realizable for this row. We denote by $L(S)$ the maximum length of a rule from $S$, and by $L(T)$ we denote the minimum value of $L(S)$ among all complete decision rule systems $S$ for $T$.

A *test for* $T$ is a subset of columns such that at the intersection with these columns any two rows with different decisions are different. A *reduct for*

$T$ is a test for $T$ for which each proper subset is not a test. It is clear that each test has a reduct as a subset. We denote by $R(T)$ the minimum cardinality of a reduct for $T$.

## 2.2 Sets of Tests, Decision Rules and Trees

In this section, we consider some results related to the structure of the set of all tests for a decision table $T$, structure of the set of decision rules which are true for $T$ and realizable for a row $r$, and the structure of decision trees for $T$.

We begin our consideration from monotone Boolean functions which will be used for description of the set of tests and the set of decision rules.

### *2.2.1 Monotone Boolean Functions*

We define a partial order $\leq$ on the set $E_2^n$ where $E_2 = \{0, 1\}$ and $n$ is a natural number. Let $\bar{\alpha} = (\alpha_1, \ldots, \alpha_n)$, $\bar{\beta} = (\beta_1, \ldots, \beta_n) \in E_2^n$. Then $\bar{\alpha} \leq \bar{\beta}$ if and only if $\alpha_i \leq \beta_i$ for $i = 1, \ldots, n$. The inequality $\bar{\alpha} < \bar{\beta}$ means that $\bar{\alpha} \leq \bar{\beta}$ and $\bar{\alpha} \neq \bar{\beta}$. Two tuples $\bar{\alpha}$ and $\bar{\beta}$ are *incomparable* if both relations $\bar{\alpha} \leq \bar{\beta}$ and $\bar{\beta} \leq \bar{\alpha}$ do not hold. A set $A \subseteq E_2^n$ is called *independent* if every two tuples from $A$ are incomparable. We omit the proofs of the following three lemmas containing well known results.

**Lemma 2.1.** *a) If $A \subseteq E_2^n$ and $A$ is an independent set then $|A| \in \{0, 1, \ldots, \binom{n}{\lfloor n/2 \rfloor}\}$.*
*b) For any $k \in \{0, 1, \ldots, \binom{n}{\lfloor n/2 \rfloor}\}$ there exists an independent set $A \subseteq E_2^n$ such that $|A| = k$.*

A function $f : E_2^n \to E_2$ is called *monotone* if for every tuples $\bar{\alpha}, \bar{\beta} \in E_2^n$ if $\bar{\alpha} \leq \bar{\beta}$ then $f(\bar{\alpha}) \leq f(\bar{\beta})$.

A tuple $\bar{\alpha} \in E_2^n$ is called an *upper zero* of the monotone function $f$ if $f(\bar{\alpha}) = 0$ and for any tuple $\bar{\beta}$ such that $\bar{\alpha} < \bar{\beta}$ we have $f(\bar{\beta}) = 1$. A tuple $\bar{\alpha} \in E_2^n$ is called a *lower unit* of the monotone function $f$ if $f(\bar{\alpha}) = 1$ and $f(\bar{\beta}) = 0$ for any tuple $\bar{\beta}$ such that $\bar{\beta} < \bar{\alpha}$.

**Lemma 2.2.** *Let $f : E_2^n \to E_2$ be a monotone function. Then*

*a) For any $\bar{\alpha} \in E_2^n$ the equality $f(\bar{\alpha}) = 1$ holds if and only if there exists a lower unit $\bar{\beta}$ of $f$ such that $\bar{\beta} \leq \bar{\alpha}$.*
*b) For any $\bar{\alpha} \in E_2^n$ the equality $f(\bar{\alpha}) = 0$ holds if and only if there exists an upper zero $\bar{\beta}$ of $f$ such that $\bar{\alpha} \leq \bar{\beta}$.*

**Lemma 2.3.** *a) For any monotone function $f : E_2^n \to E_2$ the set of lower units is an independent set.*
*b) Let $A \subseteq E_2^n$ and $A$ be an independent set. Then there exists a monotone function $f : E_2^n \to E_2$ for which the set of lower units coincides with $A$.*

### *2.2.2 Set of Tests*

Let $T$ be a decision table with $n$ columns labeled with attributes $f_1, \ldots, f_n$. There exists a one-to-one correspondence between $E_2^n$ and the set of subsets of attributes from $T$. Let $\bar{\alpha} \in E_2^n$ and $i_1, \ldots, i_m$ be numbers of digits from $\bar{\alpha}$ which are equal to 1. Then the set $\{f_{i_1}, \ldots, f_{i_m}\}$ corresponds to the tuple $\bar{\alpha}$.

Let us correspond a *characteristic function* $f_T : E_2^n \to E_2$ to the table $T$. For $\alpha \in E_2^n$ we have $f_T(\bar{\alpha}) = 1$ if and only if the set of attributes (columns) corresponding to $\bar{\alpha}$ is a test for $T$.

We omit the proof of the following simple statement.

**Lemma 2.4.** *For any decision table $T$ the function $f_T$ is a monotone function which does not equal to 0 identically and for which the set of lower units coincides with the set of tuples corresponding to reducts for the table $T$.*

**Corollary 2.5.** *For any decision table $T$ any test for $T$ contains a reduct for $T$ as a subset.*

Let us correspond a decision table $\tau(T)$ to the decision table $T$. The table $\tau(T)$ has $n$ columns labeled with attributes $f_1, \ldots, f_n$. The first row of $\tau(T)$ is filled by 1. The set of all other rows coincides with the set of all rows of the kind $l(\bar{\delta}_1, \bar{\delta}_2)$ where $\bar{\delta}_1$ and $\bar{\delta}_2$ are arbitrary rows of $T$ labeled with different decisions, and $l(\bar{\delta}_1, \bar{\delta}_2)$ is the row containing at the intersection with the column $f_i$, $i = 1, \ldots, n$, the number 0 if and only if $\bar{\delta}_1$ and $\bar{\delta}_2$ have different numbers at the intersection with the column $f_i$. The first row of $\tau(T)$ is labeled with the decision 1. All other rows are labeled with the decision 2.

We denote by $C(T)$ the decision table obtained from $\tau(T)$ by the removal all rows $\bar{\sigma}$ for each of which there exists a row $\bar{\delta}$ of the table $\tau(T)$ that is different from the first row and satisfies the inequality $\bar{\sigma} < \bar{\delta}$. The table $C(T)$ will be called the *canonical form of the table $T$*.

**Lemma 2.6.** *For any decision table $T$,*

$$f_T = f_{C(T)} \cdot$$

*Proof.* One can show that $f_T = f_{\tau(T)}$. Let us prove that $f_{\tau(T)} = f_{C(T)}$. It is not difficult to check that $f_{C(T)}(\bar{\alpha}) = 0$ if an only if there exists a row $\bar{\delta}$ of $C(T)$ labeled with the decision 2 for which $\bar{\alpha} \le \bar{\delta}$. Similar statement is true for the table $\tau(T)$.

It is clear that each row of $C(T)$ is also a row in $\tau(T)$, and equal rows in these tables are labeled with equal decisions. Therefore if $f_{\tau(T)}(\bar{\alpha}) = 1$ then $f_{C(T)}(\bar{\alpha}) = 1$.

Let $f_{C(T)}(\bar{\alpha}) = 1$. We will show that $f_{\tau(T)}(\alpha) = 1$. Let us assume the contrary. Then there exists a row $\bar{\sigma}$ from $\tau(T)$ which is labeled with the decision 2 and for which $\bar{\alpha} \le \bar{\sigma}$. From the description of $C(T)$ it follows that there exists a row $\bar{\delta}$ from $C(T)$ which is labeled with the decision 2 and for which $\bar{\sigma} \le \bar{\delta}$. But in this case $\bar{\alpha} \le \bar{\delta}$ which is impossible. Hence $f_{\tau(T)}(\alpha) = 1$ and $f_{\tau(T)} = f_{C(T)}$. □

**Lemma 2.7.** *For any decision table $T$ the set of rows of the table $C(T)$ with the exception of the first row coincides with the set of upper zeros of the function $f_T$.*

*Proof.* Let $\bar{\alpha}$ be an upper zero of the function $f_T$. Using Lemma 2.6 we obtain $f_{C(T)}(\bar{\alpha}) = 0$. Therefore there exists a row $\bar{\delta}$ in $C(T)$ which is labeled with the decision 2 and for which $\bar{\alpha} \leq \bar{\delta}$. Evidently, $f_{C(T)}(\bar{\delta}) = 0$. Therefore $f_T(\bar{\delta}) = 0$. Taking into account that $\bar{\alpha}$ is an upper zero of the function $f_T$ we conclude that the inequality $\bar{\alpha} < \bar{\delta}$ does not hold. Hence $\bar{\alpha} = \bar{\delta}$ and $\bar{\alpha}$ is a row of $C(T)$ which is labeled with the decision 2.

Let $\bar{\delta}$ be a row of $C(T)$ different from the first row. Then, evidently, $f_{C(T)}(\bar{\delta}) = 0$, and by Lemma 2.6, $f_T(\bar{\delta}) = 0$. Let $\bar{\delta} < \bar{\sigma}$. We will show that $f_T(\bar{\sigma}) = 1$. Let us assume the contrary. Then by Lemma 2.6, $f_{C(T)}(\bar{\sigma}) = 0$. Therefore there exists a row $\bar{\gamma}$ of $C(T)$ which is labeled with the decision 2 and for which $\bar{\delta} < \bar{\gamma}$. But this is impossible since any two different rows of $C(T)$ which are labeled with 2 are incomparable. Hence $f_T(\bar{\sigma}) = 1$, and $\bar{\delta}$ is an upper zero of the function $f_T$. □

We will say that two decision tables with the same number of columns are *almost equal* if the set of rows of the first table is equal to the set of rows of the second table, and equal rows in these tables are labeled with equal decisions. Almost means that corresponding columns in two tables can be labeled with different attributes.

**Proposition 2.8.** *Let $T_1$ and $T_2$ be decision tables with the same number of columns. Then $f_{T_1} = f_{T_2}$ if and only if the tables $C(T_1)$ and $C(T_2)$ are almost equal.*

*Proof.* If $f_{T_1} = f_{T_2}$ then the set of upper zeros of $f_{T_1}$ is equal to the set of upper zeros of $f_{T_2}$. Using Lemma 2.7 we conclude that the tables $C(T_1)$ and $C(T_2)$ are almost equal.

Let the tables $C(T_1)$ and $C(T_2)$ be almost equal. By Lemma 2.7, the set of upper zeros of $f_{T_1}$ is equal to the set of upper zeros of $f_{T_2}$. Using Lemma 2.2 we obtain $f_{T_1} = f_{T_2}$. □

**Theorem 2.9.** *a) For any decision table $T$ the function $f_T$ is a monotone Boolean function which does not equal to 0 identically.*
*b) For any monotone Boolean function $f : E_2^n \to E_2$ which does not equal to 0 identically there exists a decision table $T$ with $n$ columns for which $f = f_T$.*

*Proof.* a) The first part of theorem statement follows from Lemma 2.4.
b) Let $f : E_2^n \to E_2$ be a monotone Boolean function which does not equal to 0 identically, and $\{\bar{\alpha}_1, \ldots, \bar{\alpha}_m\}$ be the set of upper zeros of $f$. We consider a decision table $T$ with $n$ columns in which the first row is filled by 1, and the set of all other rows coincides with $\{\bar{\alpha}_1, \ldots, \bar{\alpha}_m\}$. The first row is labeled with the decision 1, and all other rows are labeled with the decision 2.

One can show that $C(T) = T$. Using Lemma 2.7 we conclude that the set of upper zeros of the function $f$ coincides with the set of upper zeros of the function $f_T$. From here and from Lemma 2.2 it follows that $f = f_T$. □

**Theorem 2.10.** *a) For any decision table $T$ with $n$ columns the set of tuples from $E_2^n$ corresponding to reducts for $T$ is a nonempty independent set.*
*b) For any nonempty independent subset $A$ of the set $E_2^n$ there exists a decision table $T$ with $n$ columns for which the set of tuples corresponding to reducts for $T$ coincides with $A$.*

*Proof.* The first part of theorem statement follows from Lemmas 2.2, 2.3 and 2.4. The second part of theorem statement follows from Lemmas 2.3, 2.4 and Theorem 2.9. □

**Corollary 2.11.** *a) For any decision table $T$ with $n$ columns the cardinality of the set of reducts for $T$ is a number from the set $\{1, \ldots, \binom{n}{\lfloor n/2 \rfloor}\}$.*
*b) For any $k \in \{1, \ldots, \binom{n}{\lfloor n/2 \rfloor}\}$ there exists a decision table $T$ with $n$ columns for which the number of reducts for $T$ is equal to $k$.*

Let $T$ be a decision table with $n$ columns labeled with attributes $f_1, \ldots, f_n$. It is possible to represent the function $f_T$ as a formula (conjunctive normal form) over the basis $\{\wedge, \vee\}$. We correspond to each row $\bar{\delta}$ of $C(T)$ different from the first row the disjunction $d(\bar{\delta}) = x_{i_1} \vee \ldots \vee x_{i_m}$ where $f_{i_1}, \ldots, f_{i_m}$ are all columns of $C(T)$ at the intersection with which $\bar{\delta}$ has 0. Then $f_T = \bigwedge_{\bar{\delta} \in \Delta(C(T)) \setminus \{\bar{1}\}} d(\bar{\delta})$ where $\Delta(C(T))$ is the set of rows of the table $C(T)$ and $\bar{1}$ is the first row of $C(T)$ filled by 1.

If we multiply all disjunctions and apply rules $A \vee A \wedge B = A$ and $A \wedge A = A \vee A = A$ we obtain the reduced disjunctive normal form of the function $f_T$ such that there exists a one-to-one correspondence of elementary conjunctions in this form and lower units of the functions $f_T$ (reducts for $T$): an elementary conjunction $x_{i_1} \wedge \ldots \wedge x_{i_m}$ corresponds to the lower unit of $f_T$ which has 1 only in digits $i_1, \ldots, i_m$ (corresponds to the reduct $\{f_{i_1}, \ldots, f_{i_m}\}$).

Another way for construction of a formula for the function $f_T$ is considered in Sect. 4.3.3.

*Example 2.12.* For a given decision table $T$ we construct corresponding tables $\tau(T)$ and $C(T)$—see Fig. 2.2.

We can represent the function $f_T$ as a conjunctive normal form and transform it into reduced disjunctive normal form: $f_T(x_1, x_2, x_3, x_4) = (x_2 \vee x_4) \wedge (x_3 \vee x_4) \wedge x_1 = x_2x_3x_1 \vee x_2x_4x_1 \vee x_4x_3x_1 \vee x_4x_4x_1 = x_2x_3x_1 \vee x_2x_4x_1 \vee x_4x_3x_1 \vee x_4x_1 = x_2x_3x_1 \vee x_4x_1$. Therefore the function $f_T$ has two lower units $(1, 1, 1, 0)$ and $(1, 0, 0, 1)$, and the table $T$ has two reducts $\{f_1, f_2, f_3\}$ and $\{f_1, f_4\}$.

So we have the following situation now: there is a polynomial algorithm which for a given decision table $T$ constructs its canonical form $C(T)$ and the set

$T =$

| $f_1$ | $f_2$ | $f_3$ | $f_4$ | |
|---|---|---|---|---|
| 0 | 1 | 1 | 1 | 3 |
| 0 | 0 | 1 | 0 | 2 |
| 0 | 1 | 0 | 0 | 2 |
| 1 | 1 | 0 | 0 | 3 |

$\tau(T) =$

| $f_1$ | $f_2$ | $f_3$ | $f_4$ | |
|---|---|---|---|---|
| 1 | 1 | 1 | 1 | 1 |
| 1 | 0 | 1 | 0 | 2 |
| 1 | 1 | 0 | 0 | 2 |
| 0 | 0 | 0 | 1 | 2 |
| 0 | 1 | 1 | 1 | 2 |

$C(T) =$

| $f_1$ | $f_2$ | $f_3$ | $f_4$ | |
|---|---|---|---|---|
| 1 | 1 | 1 | 1 | 1 |
| 1 | 0 | 1 | 0 | 2 |
| 1 | 1 | 0 | 0 | 2 |
| 0 | 1 | 1 | 1 | 2 |

**Fig. 2.2**

of upper zeros of the characteristic function $f_T$. If $T$ has $m$ rows then the number of upper zeros is at most $m(m-1)/2$. Based on $C(T)$ we can in polynomial time construct a formula (conjunctive normal form) over the basis $\{\wedge, \vee\}$ which represents the function $f_T$. By transformation of this formula into reduced disjunctive normal form we can find all lower units of $f_T$ and all reducts for $T$. Unfortunately, we can not guarantee that this last step will have polynomial time complexity.

*Example 2.13.* Let us consider a decision table $T$ with $m+1$ rows and $2m$ columns labeled with attributes $f_1, \ldots, f_{2m}$. The last row of $T$ is filled by 1. For $i = 1, \ldots, m$, the $i$-th row of $T$ has 0 only at the intersection with columns $f_{2i-1}$ and $f_{2i}$. The first $m$ rows of $T$ are labeled with the decision 1 and the last row is labeled with the decision 2. One can show that $f_T = (x_1 \vee x_2) \wedge (x_3 \vee x_4) \wedge \ldots \wedge (x_{2m-1} \vee x_{2m})$. This function has exactly $2^m$ lower units, and the table $T$ has exactly $2^m$ reducts.

### 2.2.3 Set of Decision Rules

Let $T$ be a decision table with $n$ columns labeled with attributes $f_1, \ldots, f_n$ and $r = (\delta_1, \ldots, \delta_n)$ be a row of $T$ labeled with a decision $d$.

We can describe the set of all decision rules over $T$ which are true for $T$ and realizable for $r$ (we will say about such rules as about rules for $T$ and $r$) with the help of *characteristic function* $f_{T,r} : E_2^n \to E$ *for* $T$ *and* $r$. Let $\bar{\alpha} \in E_2^n$ and $i_1, \ldots, i_m$ be numbers of digits from $\bar{\alpha}$ which are equal to 1. Then $f_{T,r}(\bar{\alpha}) = 1$ if and only if the rule

$$f_{i_1} = \delta_{i_1} \wedge \ldots \wedge f_{i_m} = \delta_{i_m} \to d \tag{2.1}$$

is a decision rule for $T$ and $r$. We will say that the rule (2.1) corresponds to the tuple $\bar{\alpha}$, and the tuple $\bar{\alpha}$ corresponds to the rule (2.1).

Let us correspond a decision table $T(r)$ to the table $T$. The table $T(r)$ has $n$ columns labeled with attributes $f_1, \ldots, f_n$. This table contains the row $r$ and all rows from $T$ which are labeled with decisions different from $d$. The row $r$ in $T(r)$ is labeled with the decision 1, all other rows in $T(r)$ are labeled with the decision 2. One can show that a set of attributes (columns) $\{f_{i_1}, \ldots, f_{i_m}\}$ is a test for $T(r)$ if and only if the decision rule (2.1) is a rule for $T$ and $r$. Thus $f_{T,r} = f_{T(r)}$.

We denote $C(T, r) = C(T(r))$. This table is the *canonical form for* $T$ *and* $r$. The set of rows of $C(T, r)$ with the exception of the first row coincides with the set of upper zeros of the function $f_{T,r}$ (see Lemma 2.7). Based on the table $C(T, r)$ we can represent function $f_{T,r}$ in the form of conjunctive normal form and transform this form into reduced disjunctive normal form. As a result, we obtain the set of lower units of $f_{T,r}$ which corresponds to the set of so-called irreducible decision rules for $T$ and $r$. A decision rule for $T$ and $r$ is called *irreducible* if any rule obtained from the considered one by the removal of an equality from the left-hand side is not a rule for $T$ and $r$. One can show that a set of attributes $\{f_{i_1}, \ldots, f_{i_m}\}$ is a reduct for $T(r)$ if and only if the decision rule (2.1) is an irreducible decision rule for $T$ and $r$.

**Theorem 2.14.** *a) For any decision table* $T$ *and any row* $r$ *of* $T$ *the function* $f_{T,r}$ *is a monotone Boolean function which does not equal to* 0 *identically.*
*b) For any monotone Boolean function* $f : E_2^n \to E_2$ *which does not equal to* 0 *identically there exists a decision table* $T$ *with* $n$ *columns and a row* $r$ *of* $T$ *for which* $f = f_{T,r}$.

*Proof.* a) We know that $f_{T,r} = f_{T(r)}$. From Lemma 2.4 it follows that $f_{T(r)}$ is a monotone Boolean function which does not equal to 0 identically.
b) Let $f : E_2^n \to E_2$ be a monotone Boolean function which does not equal to 0 identically, and $\{\bar{\alpha}_1, \ldots, \bar{\alpha}_m\}$ be the set of upper zeros of $f$. We consider a decision table $T$ with $n$ columns in which the first row is filled by 1 (we denote this row by $r$), and the set of all other rows coincides with $\{\bar{\alpha}_1, \ldots, \bar{\alpha}_m\}$. The first row is labeled with the decision 1 and all other rows are labeled with the decision 2.

One can show that $C(T, r) = C(T(r)) = T(r) = T$. We know that $f_{T,r} = f_{T(r)}$. So $f_T = f_{T,r}$. Using Lemma 2.7 we conclude that the set of upper zeros of $f$ coincides with the set of upper zeros of $f_T$. From here and from Lemma 2.2 it follows that $f = f_T$. Therefore $f = f_{T,r}$. □

**Theorem 2.15.** *a) For any decision table* $T$ *with* $n$ *columns and for any row* $r$ *of* $T$ *the set of tuples from* $E_2^n$ *corresponding to irreducible decision rules for* $T$ *and* $r$ *is a nonempty independent set.*
*b) For any nonempty independent subset* $A$ *of the set* $E_2^n$ *there exists a decision table* $T$ *with* $n$ *columns and row* $r$ *of* $T$ *for which the set of tuples corresponding to irreducible decision rules for* $T$ *and* $r$ *coincides with* $A$.

*Proof.* a) We know that the set of tuples corresponding to irreducible decision rules for $T$ and $r$ coincides with the set of tuples corresponding to reducts for $T(r)$. Using Theorem 2.10 we conclude that the considered set of tuples is a nonempty independent set.
b) Let $A \subseteq E_2^n$, $A \neq \emptyset$ and $A$ be independent. Using Lemma 2.3 we conclude that there exists a monotone Boolean function $f : E_2^n \to E$ for which the set of lower units coincides with $A$. From Theorem 2.14 it follows that there exists a decision table with $n$ columns and a row $r$ of $T$ for which

$f = f_{T,r}$. It is clear that the set of lower units of $f_{T,r}$ coincides with the set of tuples corresponding to irreducible decision rules for $T$ and $r$. □

**Corollary 2.16.** *a) For any decision table $T$ with $n$ columns and any row $r$ of $T$ the cardinality of the set of irreducible decision rules for $T$ and $r$ is a number from the set $\{1, 2, \ldots, \binom{n}{\lfloor n/2 \rfloor}\}$.*

*b) For any $k \in \{1, 2, \ldots, \binom{n}{\lfloor n/2 \rfloor}\}$ there exists a decision table $T$ with $n$ columns and a row $r$ of $T$ for which the number of irreducible decision rules for $T$ and $r$ is equal to $k$.*

Let $T$ be a decision table with $n$ columns labeled with attributes $f_1, \ldots, f_n$ and $r$ be a row of $T$.

As for the case of reducts, we can represent the function $f_{T,r}$ as a conjunctive normal form and transform it into reduced disjunctive normal form. As a result we obtain all irreducible decision rules for $T$ and $r$.

*Example 2.17.* Let $T$ be the decision table depicted in Fig. 2.2 and $r$ be the first row of $T$. We construct tables $T(r)$, $\tau(T(r))$ and $C(T, r) = C(T(r))$—see Fig. 2.3.

$T(r) =$

| $f_1$ | $f_2$ | $f_3$ | $f_4$ | |
|---|---|---|---|---|
| 0 | 1 | 1 | 1 | 3 |
| 0 | 0 | 1 | 0 | 2 |
| 0 | 1 | 0 | 0 | 2 |

$\tau(T(r)) =$

| $f_1$ | $f_2$ | $f_3$ | $f_4$ | |
|---|---|---|---|---|
| 1 | 1 | 1 | 1 | 1 |
| 1 | 0 | 1 | 0 | 2 |
| 1 | 1 | 0 | 0 | 2 |

$C(T, r) =$

| $f_1$ | $f_2$ | $f_3$ | $f_4$ | |
|---|---|---|---|---|
| 1 | 1 | 1 | 1 | 1 |
| 1 | 0 | 1 | 0 | 2 |
| 1 | 1 | 0 | 0 | 2 |

**Fig. 2.3**

We can represent the function $f_{T,r}$ as a conjunctive normal form and transform it into reduced disjunctive normal form: $f_{T,r}(x_1, x_2, x_3, x_4) = (x_2 \vee x_4) \wedge (x_3 \vee x_4) = x_2x_3 \vee x_2x_4 \vee x_4x_3 \vee x_4x_4 = x_2x_3 \vee x_2x_4 \vee x_4x_3 \vee x_4 = x_2x_3 \vee x_4$. Therefore the function $f_{T,r}$ has two lower units $(0, 1, 1, 0)$ and $(0, 0, 0, 1)$, and there are two irreducible decision rules for $T$ and $r$: $f_2 = 1 \wedge f_3 = 1 \rightarrow 3$ and $f_4 = 1 \rightarrow 3$.

So we have polynomial algorithms that allow us for a given decision table $T$ and row $r$ of $T$ construct the canonical form $C(T, r)$ for $T$ and $r$ and the set of upper zeros of the characteristic function $f_{T,r}$. If $T$ has $m$ rows then the number of upper zeros is at most $m - 1$. Based on $C(T, r)$ we can construct a conjunctive normal form representing $f_{T,r}$.

Also we can transform this form into the reduced disjunctive normal form and find all irreducible decision rules for $T$ and $r$. As for the case of reducts, this step can have exponential time complexity.

*Example 2.18.* Let $T$ be the decision table considered in Example 2.13 and $r$ be the last row of $T$. The table $T$ has $m + 1$ rows and $2m$ columns. One can show that $f_{T,r} = (x_1 \vee x_2) \wedge (x_3 \vee x_4) \wedge \ldots \wedge (x_{2m-1} \vee x_{2m})$, and there are exactly $2^m$ lower units of $f_{T,r}$ and $2^m$ irreducible decision rules for $T$ and $r$.

### 2.2.4 Set of Decision Trees

Decision tree is a more complicated object than test or decision rule. So we can't describe the set of decision trees in the same way as the set of tests or the set of rules. However, we can compare efficiently sets of decision trees for two decision tables with the same sets of attributes.

Let $T$ be a decision table. We denote by $DT(T)$ the set of decision trees for the table $T$. By $\varrho(T)$ we denote the set of pairs $(r, d)$ where $r$ is a row of $T$ and $d$ is the decision attached to the row $r$. We will say that two decision tables $T_1$ and $T_2$ are *equal* if $T_1$ and $T_2$ have the same number of columns labeled with the same names of attributes and $\varrho(T_1) = \varrho(T_2)$. Let $T_1$ and $T_2$ have the same number of columns. We will say that $T_1$ and $T_2$ are *consistent* if for any two pairs $(\bar{\delta}_1, d_1) \in \varrho(T_1)$ and $(\bar{\delta}_2, d_2) \in \varrho(T_2)$ from the equality $\bar{\delta}_1 = \bar{\delta}_2$ the equality $d_1 = d_2$ follows. If $T_1$ and $T_2$ are not consistent they will be called *inconsistent.*

**Theorem 2.19.** *Let $T_1$ and $T_2$ be decision tables with $n$ columns labeled with attributes $f_1, \ldots, f_n$. Then*

*a) $DT(T_1) \cap DT(T_2) \neq \emptyset$ if and only if the tables $T_1$ and $T_2$ are consistent.*
*b) $DT(T_1) \subseteq DT(T_2)$ if and only if $\varrho(T_2) \subseteq \varrho(T_1)$.*

*Proof.* a) Let $\Gamma \in DT(T_1) \cap DT(T_2)$, $(\bar{\delta}, d_1) \in \varrho(T_1)$, $(\bar{\delta}, d_2) \in \varrho(T_2)$, and $v$ be the terminal node of $\Gamma$ such that the work of $\Gamma$ for $\bar{\delta}$ finishes in $v$. Since $\Gamma \in DT(T_1)$, the node $v$ is labeled with $d_1$. Since $\Gamma \in DT(T_2)$, the node $v$ is labeled with $d_2$. Therefore $d_1 = d_2$.

Let $T_1$ and $T_2$ be consistent. Consider a decision tree $\Gamma(T_1, T_2)$ over $T_1$ which consists of $n + 1$ layers. For $i = 1, \ldots, n$, all nodes from the $i$-th layer are labeled with the attribute $f_i$. All nodes from the $(n+1)$-th layer are terminal nodes. Let $v$ be an arbitrary terminal node and $\delta_1, \ldots, \delta_n$ be numbers attached to edges in the path from the root of $\Gamma(T_1, T_2)$ to $v$. Denote $\bar{\delta} = (\delta_1, \ldots, \delta_n)$. If $\bar{\delta}$ is not a row of $T_1$ and $T_2$ then we mark $v$ with 1. If $\bar{\delta}$ is a row of $T_1$ or $T_2$ and $\bar{\delta}$ is labeled with the decision $d$ then we mark $v$ with $d$. If $\bar{\delta}$ is a row of $T_1$ and a row of $T_2$ then $\bar{\delta}$ is labeled with the same decision in $T_1$ and $T_2$ since $T_1$ and $T_2$ are consistent. It is clear that $\Gamma(T_1, T_2) \in DT(T_1) \cap DT(T_2)$.

b) Let $\varrho(T_2) \subseteq \varrho(T_1)$ and $\Gamma \in DT(T_1)$. It is clear that $\Gamma \in DT(T_2)$. Therefore $DT(T_1) \subseteq DT(T_2)$.

Let $\varrho(T_2) \not\subseteq \varrho(T_1)$. We show that $DT(T_1) \not\subseteq DT(T_2)$.

Let $T_1$ and $T_2$ be inconsistent. By proved above, $DT(T_1) \cap DT(T_2) = \emptyset$. It is clear that $DT(T_1) \neq \emptyset$. Therefore $DT(T_1) \not\subseteq DT(T_2)$.

Let $T_1$ and $T_2$ be consistent. Since $\varrho(T_2) \not\subseteq \varrho(T_1)$, there exists $\bar{\delta} = (\delta_1, \ldots, \delta_n) \in E_2^n$ and natural $d$ such that $(\bar{\delta}, d) \in \varrho(T_2)$ and $(\bar{\delta}, d) \notin \varrho(T_1)$. In the decision tree $\Gamma(T_1, T_2)$ described above, let $v$ be the terminal node such that the edges in the path from the root of $\Gamma(T_1, T_2)$ to $v$ are labeled with the numbers $\delta_1, \ldots, \delta_n$. In $\Gamma(T_1, T_2)$ the node $v$ is labeled with $d$. Instead of $d$ we mark $v$ with $d + 1$. We denote the obtained decision tree by $\Gamma$. It is clear that $\Gamma \in DT(T_1)$ and $\Gamma \notin DT(T_2)$. Therefore $DT(T_1) \nsubseteq DT(T_2)$. □

**Corollary 2.20.** *Let $T_1$ and $T_2$ be decision tables with $n$ columns labeled with attributes $f_1, \ldots, f_n$. Then $DT(T_1) = DT(T_2)$ if and only if $\varrho(T_1) = \varrho(T_2)$.*

It is interesting to compare Corollary 2.20 and Proposition 2.8. From this proposition it follows that the set of tests for $T_1$ is equal to the set of tests for $T_2$ if and only if $\varrho(C(T_1)) = \varrho(C(T_2))$. This condition is essentially weaker than the condition $\varrho(T_1) = \varrho(T_2)$: it is possible that for very different decision tables $T_1$ and $T_2$ the canonical forms $C(T_1)$ and $C(T_2)$ are equal.

*Example 2.21.* Let us consider four decision tables (see Fig. 2.4).

$T_1 =$

| $f_1$ | $f_2$ | |
|---|---|---|
| 0 | 1 | 1 |
| 1 | 0 | 2 |

$T_2 =$

| $f_1$ | $f_2$ | |
|---|---|---|
| 1 | 0 | 2 |
| 0 | 1 | 1 |

$T_3 =$

| $f_1$ | $f_2$ | |
|---|---|---|
| 0 | 1 | 1 |
| 1 | 0 | 2 |
| 0 | 0 | 3 |

$T_4 =$

| $f_1$ | $f_2$ | |
|---|---|---|
| 0 | 1 | 1 |
| 1 | 0 | 2 |
| 0 | 0 | 2 |

**Fig. 2.4**

We have $DT(T_1) = DT(T_2)$, $DT(T_3) \subset DT(T_1)$, $DT(T_4) \subset DT(T_1)$ and $DT(T_3) \cap DT(T_4) = \emptyset$.

For any decision table $T$, the number of decision trees for $T$ is infinite. However, often we can narrow down the consideration to irreducible decision trees for $T$ which number is finite.

Let $T$ be a decision table with $n$ columns labeled with attributes $f_1, \ldots, f_n$. We denote by $E(T)$ the set of attributes (columns of $T$) each of which contains different numbers. Let $\Gamma$ be a decision tree over $T$ and $v$ be a node of $\Gamma$. Let in the path from the root of $\Gamma$ to $v$ nodes be labeled with attributes $f_{i_1}, \ldots, f_{i_m}$ and edges be labeled with numbers $\delta_1, \ldots, \delta_m$. Denote by $T(v)$ the subtable of $T$ which consists of rows that at the intersection with columns $f_{i_1}, \ldots, f_{i_m}$ have numbers $\delta_1, \ldots, \delta_m$.

We will say that a decision tree $\Gamma$ for $T$ is *irreducible* if any node $v$ of $\Gamma$ satisfies the following conditions:

1. If all rows of the subtable $T(v)$ are labeled with the same decision $d$ then $v$ is a terminal node of $\Gamma$ labeled with $d$.
2. If there are rows of $T(v)$ labeled with different decisions then $v$ is a working node labeled with an attribute from $E(T(v))$.

**Proposition 2.22.** *Let $T$ be a decision table with $n$ columns labeled with attributes $f_1, \ldots, f_n$. Then there exists an irreducible decision tree $\Gamma$ for $T$ such that $h(\Gamma) = h(T)$.*

*Proof.* Let $D$ be a decision tree for $T$ such that $h(D) = h(T)$. We will modify the tree $D$ in order to obtain an irreducible decision tree for $T$. The algorithm considers working nodes of the tree $D$ sequentially beginning with the root. Let $v$ be the current node and $f_i$ be an attribute attached to the node $v$. The algorithm tries to apply the following rules to the node $v$.

1. Let all rows of $T(v)$ be labeled with the same decision $d$. Then remove all descendants of $v$ and label $v$ with $d$ instead of $f_i$.
2. Let $f_i \notin E(T(v))$ and the column $f_i$ in the table $T(v)$ contain only one number $a$. Denote by $\Gamma_a$ the tree which root is the end of the edge started in $v$ and labeled with $a$. Then substitute the subtree whose root is $v$ to $\Gamma_a$.

We denote the obtained decision tree by $\Gamma$. One can show that $\Gamma$ is an irreducible decision tree for $T$ and $h(\Gamma) \leq h(D)$. Therefore $h(\Gamma) = h(T)$. □

Note that the algorithm described in the proof of Proposition 2.22 can be applied to an arbitrary decision table $T$ and an arbitrary decision tree $\Gamma$ for $T$. As a result we obtain an irreducible decision tree $\Gamma'$ for $T$ such that $h(\Gamma') \leq h(\Gamma)$.

## 2.3 Relationships among Decision Trees, Rules and Tests

**Theorem 2.23.** *Let $T$ be a decision table with $n$ columns labeled with attributes $f_1, \ldots, f_n$.*

1. *If $\Gamma$ is a decision tree for $T$ then the set of attributes attached to working nodes of $\Gamma$ is a test for the table $T$.*
2. *Let $F = \{f_{i_1}, \ldots, f_{i_m}\}$ be a test for $T$. Then there exists a decision tree $\Gamma$ for $T$ which uses only attributes from $F$ and for which $h(\Gamma) = m$.*

*Proof.* 1. Let $r_1$ and $r_2$ be two rows from $T$ with different decisions. Since $\Gamma$ is a decision tree for $T$ then the work of $\Gamma$ for $r_1$ and $r_2$ finishes in different terminal nodes. Therefore there exists an attribute $f_i$ attached to a working node of $\Gamma$ such that $r_1$ and $r_2$ are different in the column $f_i$. Since $r_1$ and $r_2$ is an arbitrary pair of rows with different decisions, we obtain that the set of attributes attached to working nodes of $\Gamma$ is a test for the table $T$.

2. Let $\{f_{i_1}, \ldots, f_{i_m}\}$ be a test for $T$. Consider a decision tree $\Gamma$ which consists of $m + 1$ layers. For $j = 1, \ldots, m$, all nodes on the $j$-th layer are labeled with the attribute $f_{i_j}$. All nodes from the $(m + 1)$-th layer are terminal nodes. Let $v$ be an arbitrary terminal node. If there is no row of $T$ for which the work of $\Gamma$ finishes in this node then we mark $v$ with the number 1. Let there exist rows of $T$ for which the work of $\Gamma$ finishes in $v$. Since $\{f_{i_1}, \ldots, f_{i_m}\}$ is a test for $T$, all these rows are labeled with the same

decision. We mark the node $v$ with this decision. It is clear that $\Gamma$ is a decision tree for $T$ which uses only attributes from $\{f_{i_1}, \ldots, f_{i_m}\}$ and which depth is equal to $m$. □

**Corollary 2.24.** *Let $T$ be a decision table. Then $h(T) \leq R(T)$.*

**Theorem 2.25.** *Let $T$ be a decision table with $n$ columns labeled with attributes $f_1, \ldots, f_n$.*

1. *If $S$ is a complete system of decision rules for $T$ then the set of attributes from rules in $S$ is a test for $T$.*
2. *If $F = \{f_{i_1}, \ldots, f_{i_m}\}$ is a test for $T$ then there exists a complete system $S$ of decision rules for $T$ which uses only attributes from $F$ and for which $L(S) = m$.*

*Proof.* 1. Let $S$ be a complete system of decision rules for $T$, and $r_1$, $r_2$ be two rows of $T$ with different decisions. Then there exists a rule from $S$ which is realizable for $r_1$ and is not realizable for $r_2$. It means that there is an attribute $f_i$ on the left-hand side of the considered rule such that $r_1$ and $r_2$ are different in the column $f_i$. Since $r_1$ and $r_2$ is an arbitrary pair of rows with different decisions, we obtain that the set of attributes from rules in $S$ is a test for $T$.

2. Let $F = \{f_{i_1}, \ldots, f_{i_m}\}$ be a test for $T$. Consider a decision rule system $S$ which contains all rules of the kind

$$f_{i_1} = b_1 \wedge \ldots \wedge f_{i_m} = b_m \rightarrow t$$

for each of which there exists a row $r$ of $T$ such that the considered rule is realizable for $r$, and $r$ is labeled with the decision $t$. Since $F$ is a test for $T$, the considered rule is true for $T$. Therefore $S$ is a complete decision rule system for $T$ and $L(S) = m$. □

**Corollary 2.26.** *$L(T) \leq R(T)$.*

Let $\Gamma$ be a decision tree for $T$ and $\tau$ be a path in $\Gamma$ from the root to a terminal node in which working nodes are labeled with attributes $f_{i_1}, \ldots, f_{i_m}$, edges are labeled with numbers $b_1, \ldots, b_m$, and the terminal node of $\tau$ is labeled with the decision $t$. We correspond to $\tau$ the decision rule rule$(\tau)$

$$f_{i_1} = b_1 \wedge \ldots \wedge f_{i_m} = b_m \rightarrow t\,.$$

**Theorem 2.27.** *Let $\Gamma$ be a decision tree for $T$, and $S$ be the set of decision rules corresponding to paths in $\Gamma$ from the root to terminal nodes. Then $S$ is a complete system of decision rules for $T$ and $L(S) = h(\Gamma)$.*

*Proof.* Since $\Gamma$ is a decision tree for $T$, for each row $r$ of $T$ there exists a path $\tau$ from the root to a terminal node $v$ of $\Gamma$ such that the work of $\Gamma$ for $r$ finishes in $v$, and $v$ is labeled with the decision $t$ attached to $r$. It is clear that $rule(\tau)$ is realizable for $r$. It is clear also that for each row $r'$ of $T$, such that $rule(\tau)$ is realizable for $r'$, the row $r'$ is labeled with the decision $t$. So, $rule(\tau)$ is true for $T$. Thus $S$ is a complete decision rule system for $T$. It is clear that the length of $rule(\tau)$ is equal to the length of the path $\tau$. Therefore $L(S) = h(\Gamma)$. □

**Corollary 2.28.** $L(T) \leq h(T)$.

## 2.4 Conclusions

The chapter is devoted to the study of the sets of tests, decision rules and trees, and relationships among these objects.

We can write efficiently formulas for characteristic functions which describe sets of tests and rules. We can find efficiently sets of upper zeros for these functions (it can be useful for design of lazy learning algorithms). However, there are no polynomial algorithms for construction of the set of lower units for characteristic functions (in the case of tests, lower units correspond to reducts, and in the case of rules—to irreducible decision rules). We can compare efficiently sets of decision trees for decision tables with the same names of attributes.

We studied relationships among decision trees, rules and tests which allow us to work effectively with bounds on complexity and algorithms for construction of rules, tests and trees.

The results considered in this chapter (with the exception of results for decision rules and decision rule systems) were published in methodical developments [38, 39] for the course Test Theory and its Applications—a predecessor of the course Combinatorial Machine Learning.

# 3

# Bounds on Complexity of Tests, Decision Rules and Trees

In this chapter, we continue the consideration of decision tables with one-valued decisions. We study bounds on complexity for decision trees, rules, and tests.

The chapter consists of three sections. In Sect. 3.1, we investigate lower bounds on the depth of decision trees, cardinality of tests and length of decision rules.

Section 3.2 is devoted to the consideration of upper bounds on the minimum cardinality of tests and minimum depth of decision trees. These bounds can be used also as upper bounds on the minimum length of decision rules.

Section 3.3 contains conclusions.

## 3.1 Lower Bounds

From Corollaries 2.24 and 2.28 it follows that $L(T) \leq h(T) \leq R(T)$. So each lower bound on $L(T)$ is a lower bound on $h(T)$ and $R(T)$, and each lower bound on $h(T)$ is also a lower bound on $R(T)$.

Let us consider now some lower bounds on the value $h(T)$ and, consequently, on the value $R(T)$.

We denote by $D(T)$ the number of different decisions in a decision table $T$.

**Theorem 3.1.** *Let $T$ be a nonempty decision table. Then*

$$h(T) \geq \log_2 D(T) .$$

*Proof.* Let $\Gamma$ be a decision tree for $T$ such that $h(\Gamma) = h(T)$. We denote by $L_t(\Gamma)$ the number of terminal nodes in $\Gamma$. It is clear that $L_t(\Gamma) \geq D(T)$. One can show that $L_t(\Gamma) \leq 2^{h(\Gamma)}$. Therefore $2^{h(\Gamma)} \geq D(T)$ and $h(\Gamma) \geq \log_2 D(T)$. Thus, $h(T) \geq \log_2 D(T)$. □

M. Moshkov and B. Zielosko: Combinatorial Machine Learning, SCI 360, pp. 37–46.
springerlink.com © Springer-Verlag Berlin Heidelberg 2011

**Theorem 3.2.** *Let $T$ be a decision table. Then*

$$h(T) \geq \log_2(R(T)+1) .$$

*Proof.* Let $\Gamma$ be a decision tree for $T$ such that $h(\Gamma) = h(T)$. We denote by $L_w(\Gamma)$ the number of working nodes in $\Gamma$. From Theorem 2.23 it follows that the set of attributes attached to working nodes of $\Gamma$ is a test for $T$. Therefore $L_w(\Gamma) \geq R(T)$. One can show that $L_w(\Gamma) \leq 1+2+\ldots+2^{h(\Gamma)-1} = 2^{h(\Gamma)}-1$. Therefore $2^{h(\Gamma)} - 1 \geq R(T)$, $2^{h(\Gamma)} \geq R(T)+1$ and $h(\Gamma) \geq \log_2(R(T)+1)$. Since $h(\Gamma) = h(T)$ we obtain $h(T) \geq \log_2(R(T)+1)$. □

*Example 3.3.* Let us consider the decision table $T$ depicted in Fig. 3.1.

$T=$

| $f_1$ | $f_2$ | $f_3$ | |
|---|---|---|---|
| 1 | 1 | 1 | 1 |
| 0 | 1 | 0 | 2 |
| 1 | 1 | 0 | 2 |
| 0 | 0 | 1 | 3 |
| 1 | 0 | 0 | 3 |

**Fig. 3.1**

For this table $D(T) = 3$. Using Theorem 3.1 we obtain $h(T) \geq \log_2 3$. Therefore $h(T) \geq 2$.

One can show that this table has exactly two tests: $\{f_1, f_2, f_3\}$ and $\{f_2, f_3\}$. Therefore $R(T) = 2$. Using Theorem 3.2 we obtain $h(T) \geq \log_2 3$ and $h(T) \geq 2$.

In fact, $h(T) = 2$. A decision tree for the table $T$ which depth is equal to 2 is depicted in Fig. 3.2.

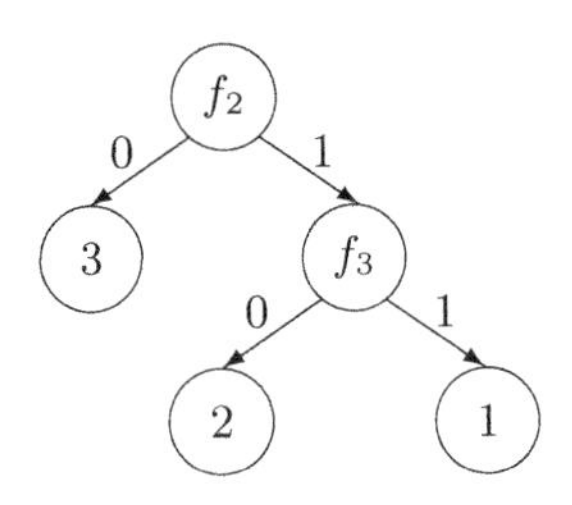


**Fig. 3.2**

Let $T$ be a decision table with $n$ columns which are labeled with attributes $f_1, \ldots, f_n$. A *subtable* of the table $T$ is a table obtained from $T$ by removal some rows. Let $\{f_{i_1}, \ldots, f_{i_m}\} \in \{f_1, \ldots, f_n\}$ and $\delta_1, \ldots, \delta_m \in \{0,1\}$. We denote by $T(f_{i_1}, \delta_1) \ldots (f_{i_m}, \delta_m)$ the subtable of the table $T$ which consists of rows that at the intersection with columns $f_{i_1}, \ldots, f_{i_m}$ have numbers $\delta_1, \ldots, \delta_m$.

We will say that $T$ is a *degenerate* table if $T$ does not have rows or all rows of $T$ are labeled with the same decision.

We define now a parameter $M(T)$ of the table $T$. If $T$ is degenerate then $M(T) = 0$. Let $T$ be nondegenerate. Let $\bar{\delta} = (\delta_1, \ldots, \delta_n) \in \{0,1\}^n$. We denote by $M(T, \bar{\delta})$ the minimum natural $m$ such that there exist $f_{i_1}, \ldots, f_{i_m} \in \{f_1, \ldots, f_n\}$ for which $T(f_{i_1}, \delta_{i_1}) \ldots (f_{i_m}, \delta_{i_m})$ is a degenerate table. Then $M(T) = \max\{M(T, \bar{\delta}) : \bar{\delta} \in \{0,1\}^n\}$.

We consider one more definition of the parameter $M(T, \bar{\delta})$. If $\bar{\delta}$ is a row of $T$ then $M(T, \bar{\delta})$ is the minimum number of columns on which $\bar{\delta}$ is different from all rows with other decisions. Let $\bar{\delta}$ be not a row of $T$. Then $M(T, \bar{\delta})$ is the minimum number of columns on which $\bar{\delta}$ is different from all rows of $T$ with the exception, possibly, of some rows with the same decision.

**Lemma 3.4.** *Let $T$ be a decision table and $T'$ be a subtable of $T$. Then*

$$M(T') \leq M(T) .$$

*Proof.* Let $T$ have $n$ columns labeled with attributes $f_1, ..., f_n$. Let $f_{i_1}, ..., f_{i_m} \in \{f_1, ..., f_n\}$ and $\delta_1, ..., \delta_m \in \{0,1\}$. If $T(f_{i_1}, \delta_1)...(f_{i_m}, \delta_m)$ is a degenerate table then $T'(f_{i_1}, \delta_1)...(f_{i_m}, \delta_m)$ is a degenerate table too. From here and from the definition of parameter $M$ the statement of lemma follows. □

*Example 3.5.* Let us find the value $M(T)$ for the decision table $T$ depicted in Fig. 3.1. To this end we find the value $M(T, \bar{\delta})$ for each $\bar{\delta} \in \{0,1\}^3$. We obtain (see Fig. 3.3) that $M(T) = 2$.

| | $f_1$ | $f_2$ | $f_3$ | | $M(T, \bar{\delta})$ |
|---|---|---|---|---|---|
| $T=$ | 1 | 1 | 1 | 1 | 2 |
| | 0 | 1 | 0 | 2 | 2 |
| | 1 | 1 | 0 | 2 | 2 |
| | 0 | 0 | 1 | 3 | 1 |
| | 1 | 0 | 0 | 3 | 1 |
| | 0 | 0 | 0 | | 1 |
| | 1 | 0 | 1 | | 1 |
| | 0 | 1 | 1 | | 2 |

**Fig. 3.3**

**Theorem 3.6.** *Let $T$ be a decision table. Then*

$$h(T) \geq M(T) .$$

*Proof.* If $T$ is a degenerate table then $h(T) = 0$ and $M(T) = 0$. Let $T$ be a nondegenerate table having $n$ columns labeled with attributes $f_1, \ldots, f_n$. Let $\Gamma$ be a decision tree for the table $T$ such that $h(\Gamma) = h(T)$. Let $\bar{\delta} =$

$(\delta_1, \ldots, \delta_n) \in \{0,1\}^n$ be a $n$-tuple for which $M(T, \bar{\delta}) = M(T)$. We consider a path $\tau = v_1, d_1, \ldots, v_m, d_m, v_{m+1}$ from the root $v_1$ to a terminal node $v_{m+1}$ in $\Gamma$ which satisfies the following condition: if nodes $v_1, \ldots, v_m$ are labeled with attributes $f_{i_1}, \ldots, f_{i_m}$ then edges $d_1, \ldots, d_m$ are labeled with numbers $\delta_{i_1}, \ldots, \delta_{i_m}$. Denote $T' = T(f_{i_1}, \delta_{i_1}) \ldots (f_{i_m}, \delta_{i_m})$. It is clear that the set of rows of $T'$ coincides with the set of rows of $T$ for which the work of $\Gamma$ finishes in the terminal node $v_{m+1}$. Since $\Gamma$ is a decision tree for the table $T$, the subtable $T'$ is a degenerate table. Therefore $m \geq M(T, \bar{\delta})$ and $h(\Gamma) \geq M(T, \bar{\delta})$. Since $h(\Gamma) = h(T)$ and $M(T, \bar{\delta}) = M(T)$, we have $h(T) \geq M(T)$. □

The following example helps us to understand why in the definition of $M(T)$ we use not only rows of $T$ but also tuples $\bar{\delta}$ which are not rows of $T$.

*Example 3.7.* Let us consider a decision table $T$ with $n$ columns and $n$ rows labeled with decisions $1, \ldots, n$. For $i = 1, \ldots, n$, the $i$-th row has 1 only at the intersection with the column $f_i$.

It is clear that for any row $\bar{\delta}$ of $T$ the equality $M(T, \bar{\delta}) = 1$ holds. Let us consider the tuple $\bar{0} = (0, \ldots, 0) \in \{0,1\}^n$. This tuple is not a row of $T$. One can show that $M(T, \bar{0}) = n - 1$.

We have now three lower bounds. Unfortunately, sometimes each of these bounds is not an exact bound. What to do in this case?

*Example 3.8.* Let us consider the problem of computation of the function $f(x, y, z) = xy \vee xz \vee yz$ with the help of decision trees using values of variables $x$, $y$ and $z$. The corresponding decision table $T$ is depicted in Fig. 3.4. Our aim is to evaluate the value of $h(T)$. Note that the value of the

$T =$

| $x$ | $y$ | $z$ | |
|---|---|---|---|
| 0 | 0 | 0 | 0 |
| 0 | 0 | 1 | 0 |
| 0 | 1 | 0 | 0 |
| 0 | 1 | 1 | 1 |
| 1 | 0 | 0 | 0 |
| 1 | 0 | 1 | 1 |
| 1 | 1 | 0 | 1 |
| 1 | 1 | 1 | 1 |

**Fig. 3.4**

considered function (the function of voting) on a tuple $(\delta_1, \delta_2, \delta_3)$ is equal to 0 if the number of 0 among $\delta_1$, $\delta_2$, and $\delta_3$ is maximum, and it is equal to 1 if the number of 1 among $\delta_1$, $\delta_2$, and $\delta_3$ is maximum.

It is clear that $D(T) = 2$. From Theorem 3.1 it follows that $h(T) \geq 1$. One can show that $R(T) = 3$. From Theorem 3.2 it follows that $h(T) \geq 2$. It is not difficult to see that for any $\bar{\delta} \in \{0,1\}^3$ the inequality $M(T, \bar{\delta}) \geq 2$ holds.

On the other hand, it is easy to notice that $M(T, \bar{\delta}) \leq 2$ for any $\bar{\delta} \in \{0,1\}^3$. Hence, $M(T) = 2$ and from Theorem 3.6 it follows that $h(T) \geq 2$. Thus, we have the following lower bound: $h(T) \geq 2$. But it is impossible to construct a decision tree for $T$ which depth is equal to 2.

What to do? We should find a way to obtain exact lower bound. In the considered case we can use the following reasoning: if the first question is, for example, about the value of $x$, then we will answer that $x = 0$; if the second question is, for example, about the value of $y$, then we will answer that $y = 1$. Thus, it will be necessary to ask the third question about the value of $z$.

In some sense we are saying about a *strategy of the first player* in the game which is *modified* in the following way: the first player does not choose a row at the beginning of the game, but at least one row must satisfy his answers on questions of the second player.

A strategy of the first player is depicted in Fig. 3.5.

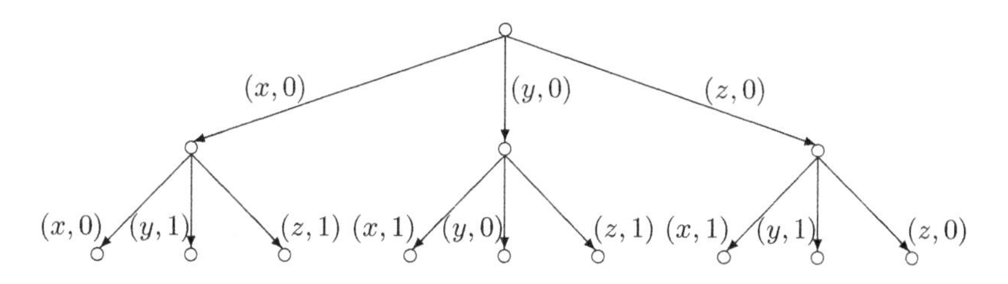


**Fig. 3.5**

We see that if the first player uses this strategy, the second player after any two questions will localize the considered row in a table which is not degenerate. Therefore he should make additional third step, and hence, $h(T) \geq 3$. A decision tree for $T$ which depth is equal to 3 is depicted in Fig. 3.6.

Now we consider the notion of strategy of the first player more formally. We will say about so-called proof-trees.

Let $T$ be a decision table with $n$ columns labeled with attributes $f_1, \ldots, f_n$, $m$ be a natural number, and $m \leq n$.

A $(T, m)$*-proof-tree* is a finite directed tree $G$ with the root in which the length of each path from the root to a terminal node is equal to $m - 1$. Nodes of this tree are not labeled. In each nonterminal node exactly $n$ edges start. These edges are labeled with pairs of the kind $(f_1, \delta_1), \ldots, (f_n, \delta_n)$ respectively where $\delta_1, \ldots, \delta_n \in \{0, 1\}$. For example, in Fig. 3.5 a $(T, 3)$-proof-tree is depicted, where $T$ is the table depicted in Fig. 3.4.

Let $v$ be an arbitrary terminal node of $G$ and $(f_{i_1}, \delta_1), \ldots, (f_{i_{m-1}}, \delta_{m-1})$ be pairs attached to edges in the path from the root of $G$ to the terminal node $v$. Denote $T(v) = T(f_{i_1}, \delta_1) \ldots (f_{i_{m-1}}, \delta_{m-1})$.

We will say that $G$ is a *proof-tree for the bound* $h(T) \geq m$ if for any terminal node $v$ the subtable $T(v)$ is not degenerate.

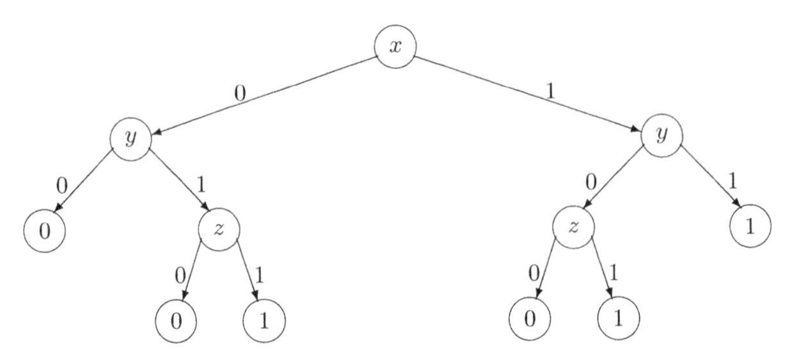


**Fig. 3.6**

**Theorem 3.9.** *Let $T$ be a nondegenerate decision table with $n$ columns and $m$ be a natural number such that $m \leq n$. Then a proof-tree for the bound $h(T) \geq m$ exists if and only if the inequality $h(T) \geq m$ holds.*

*Proof.* Let columns of $T$ be labeled with attributes $f_1, \ldots, f_n$.

Let $G$ be a proof-tree for the bound $h(T) \geq m$. We prove that $h(T) \geq m$. Let $\Gamma$ be a decision tree for $T$ such that $h(\Gamma) = h(T)$. Choose a path in $\Gamma$ from the root to some node, and a path in $G$ from the root to a terminal node in the following way. Let the root of $\Gamma$ be labeled with the attribute $f_{i_1}$. We find an edge which starts in the root of $G$ and is labeled with a pair $(f_{i_1}, \delta_1)$. We pass along this edge in the tree $G$ and pass along the edge labeled with $\delta_1$ in the tree $\Gamma$. Then we will repeat the considered procedure until we come in the tree $G$ to a terminal node $v$. In the same time we will come to a node $w$ of the tree $\Gamma$. It is clear that $T(v)$ coincides with the subtable of $T$ consisting of rows for which during the work of $\Gamma$ we pass through the node $w$. Since $T(v)$ is a not a degenerate table, $w$ is not a terminal node. Therefore the depth of $\Gamma$ is at least $m$. Since $h(\Gamma) = h(T)$, we obtain $h(T) \geq m$.

Let $h(T) \geq m$. We prove by induction on $m$ that there exists a proof-tree for the bound $h(T) \geq m$. Let $m = 1$. Then in the capacity of such proof-tree we can take the tree which consists of exactly one node. Let us assume that for some $m \geq 1$ for each decision table $T$ with $h(T) \geq m$ there exists a proof-tree for the bound $h(T) \geq m$. Let $T$ be a decision table for which $h(T) \geq m + 1$. We show that there exists a proof-tree for the bound $h(T) \geq m + 1$. Let $i \in \{1, \ldots, n\}$. It is clear that there exists $\delta_i \in \{0, 1\}$ such that $h(T(f_i, \delta_i)) \geq m$ (in the opposite case, $h(T) \leq m$ which is impossible). Using inductive hypothesis we obtain that for the table $T(f_i, \delta_i)$ there exists a proof-tree $G_i$ for the bound $h(T(f_i, \delta_i)) \geq m$. Let us construct a proof-tree $G$. In the root of $G$, $n$ edges start. These edges enter the roots of the trees $G_1, \ldots, G_n$ and are labeled with pairs $(f_1, \delta_1), \ldots, (f_n, \delta_n)$. One can show that $G$ is a proof-tree for the bound $h(T) \geq m + 1$. □

*Example 3.10.* Let us consider the decision table $T$ depicted in Fig. 3.1. We know that $h(T) = 2$ (see Example 3.3). We construct a proof-tree for the bound $h(T) \geq 2$. To this end we must find for each $i \in \{1, 2, 3\}$ a number $\delta_i \in \{0, 1\}$ such that $T(f_i, \delta_i)$ is a nondegenerate table.

It is clear that $T(f_1, 1)$, $T(f_2, 1)$, and $T(f_3, 1)$ are nondegenerate tables. Corresponding proof-tree for the bound $h(T) \geq 2$ is depicted in Fig. 3.7.

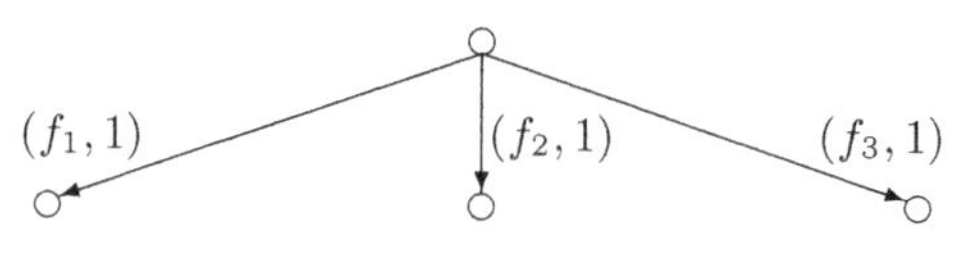


**Fig. 3.7**

We can find exact formula for the value $L(T)$ in terms of parameters $M(T, \bar{\delta})$.

**Theorem 3.11.** *Let $T$ be a decision table and $\Delta(T)$ be the set of rows of $T$. Then $L(T, \bar{\delta}) = M(T, \bar{\delta})$ for any row $\bar{\delta} \in \Delta(T)$ and $L(T) = \max\{M(T, \bar{\delta}) : \bar{\delta} \in \Delta(T)\}$.*

*Proof.* Let $\bar{\delta}$ be a row of $T$. We know that $M(T, \bar{\delta})$ is the minimum number of columns on which $\bar{\delta}$ is different from all rows with other decisions. It is clear that $M(T, \bar{\delta})$ is equal to the minimum length $L(T, \bar{\delta})$ of a decision rule which is realizable for $\bar{\delta}$ and true for $T$. Therefore $L(T) = \max_{\bar{\delta} \in \Delta(T)} M(T, \bar{\delta})$. □

**Corollary 3.12.** $L(T) \leq M(T) \leq h(T) \leq R(T)$.

*Example 3.13.* For the table $T_1$ depicted in Fig. 3.1 we have $\max\{M(T_1, \bar{\delta}) : \bar{\delta} \in \Delta(T_1)\} = 2$ (see Example 3.5). Therefore $L(T_1) = 2$.

For the decision table $T_2$ depicted in Fig. 3.4 we have $\max\{M(T_2, \bar{\delta}) : \bar{\delta} \in \Delta(T_2)\} = 2$ (see Example 3.8). Therefore $L(T_2) = 2$.

## 3.2 Upper Bounds

First, we consider an upper bound on the value $R(T)$. It will be also an upper bound on the values $L(T)$ and $h(T)$. We denote by $N(T)$ the number of rows in the table $T$.

**Theorem 3.14.** *Let $T$ be a decision table. Then*

$$R(T) \leq N(T) - 1 .$$

*Proof.* We prove this inequality by induction on $N(T)$. If $N(T) = 1$ then, evidently, $R(T) = 0$ since there are no pairs of rows with different decisions. Let $m \geq 1$ and for any decision table $T$ with $N(T) \leq m$ the inequality

$R(T) \le N(T) - 1$ holds. Let $T$ be a decision table with $N(T) = m + 1$. We prove that $R(T) \le m$. Since $T$ has at least two rows and rows of $T$ are pairwise different, there exists a column $f_i$ of $T$ which has both 0 and 1.

Let us consider subtables $T(f_i, 0)$ and $T(f_i, 1)$. It is clear that each of these subtables has at most $m$ rows. Using induction hypothesis we conclude that for $\delta \in \{0, 1\}$, there exists a test $B_\delta$ for the table $T(f_i, \delta)$ such that $|B_\delta| \le N(T(f_i, \delta)) - 1$. Denote $B = \{f_i\} \cup B_0 \cup B_1$. It is clear that $B$ is a test for the table $T$ and $|B| \le 1 + N(T(f_i, 0)) - 1 + N(T(f_i, 1)) - 1$.

Since $N(T) = N(T(f_i, 0)) + N(T(f_i, 1))$, we have $|B| \le N(T) - 1$. Therefore $R(T) \le N(T) - 1$. □

*Example 3.15.* Let $n$ be a natural number. We consider a decision table $T_n$ which contains $n$ columns labeled with conditional attributes $f_1, \ldots, f_n$ and $n + 1$ rows. For $i = 1, \ldots, n$, the $i$-th row has 1 only at the intersection with the column $f_i$. This row is labeled with the decision 1. The last $(n + 1)$-th row is filled by 0 only and is labeled with the decision 2. One can show that $N(T_n) = n + 1$ and $R(T_n) = n$. Thus, the bound from Theorem 3.14 is unimprovable in the general case.

*Example 3.16.* Let $T$ be the table depicted in Fig. 3.1. We know (see Example 3.3) that $R(T) = 2$. Theorem 3.14 gives us the upper bound $R(T) \le 4$.

Now, we consider an upper bound on the value $h(T)$. It will be also an upper bound on the value $L(T)$.

Let our decision table $T$ have a column $f_i$ in which the number of 0 is equal to the number of 1. Then after computation of $f_i$ value we localize the considered row in a subtable which has one-half rows of the table $T$. If at every step such attribute is found, then we will construct a decision tree which depth is about $\log_2 N(T)$.

Of course, very often we have no such attributes. What to do in this case? We will try to find a set of attributes $\{f_{i_1}, \ldots, f_{i_m}\}$ such that if we compute values of attributes $f_{i_1}, \ldots, f_{i_m}$ then we either will know the decision corresponding to the considered row, or localize the considered row in a subtable, which has at most one-half of rows of the table $T$.

**Theorem 3.17.** *Let $T$ be a decision table. Then*

$$h(T) \le M(T) \log_2 N(T) .$$

*Proof.* Let $T$ be a degenerate table. Then $h(T) = 0$, $M(T) = 0$ and the considered inequality holds. Let $T$ be a nondegenerate table with $n$ columns which are labeled with attributes $f_1, \ldots, f_n$. For $i = 1, \ldots, n$, let $\sigma_i$ be a number from $\{0, 1\}$ such that

$$N(T(f_i, \sigma_i)) = \max\{N(T(f_i, 0)), N(T(f_i, 1))\} .$$

Then there exist attributes $f_{i_1}, \ldots, f_{i_m} \in \{f_1, \ldots, f_n\}$ such that $m \le M(T)$ and $T(f_{i_1}, \sigma_{i_1}) \ldots (f_{i_m}, \sigma_{i_m})$ is a degenerate table.

Now we begin to describe the work of a decision tree $\Gamma$ on a row $r$ of the decision table $T$. First, we find sequentially values of the attributes $f_{i_1}, \ldots, f_{i_m}$ on the considered row. If $f_{i_1} = \sigma_{i_1}, \ldots, f_{i_m} = \sigma_{i_m}$ then our row is localized in the degenerate table $T(f_{i_1}, \sigma_{i_1}) \ldots (f_{i_m}, \sigma_{i_m})$. So we know the decision attached to this row. Let now there exist $k \in \{1, \ldots, m\}$ such that $f_{i_1} = \sigma_{i_1}, \ldots, f_{i_{k-1}} = \sigma_{i_{k-1}}$ and $f_{i_k} \neq \sigma_{i_k}$. In this case, the considered row is localized in the subtable

$$T' = T(f_{i_1}, \sigma_{i_1}) \ldots (f_{i_{k-1}}, \sigma_{i_{k-1}})(f_{i_k}, \neg\sigma_{i_k})$$

where $\neg\sigma = 0$ if $\sigma = 1$, and $\neg\sigma = 1$ if $\sigma = 0$. Since

$$N(T(f_{i_k}, \sigma_{i_k})) \geq N(T(f_{i_k}, \neg\sigma_{i_k}))$$

and $N(T) = N(T(f_{i_k}, \sigma_{i_k})) + N(T(f_{i_k}, \neg\sigma_{i_k}))$, we obtain $N(T(f_{i_k}, \neg\sigma_{i_k})) \leq N(T)/2$ and $N(T') \leq N(T)/2$.

Later the tree $\Gamma$ works similarly but instead of the table $T$ we will consider its subtable $T'$. From Lemma 3.4 it follows that $M(T') \leq M(T)$.

The process described above will be called a big step of the decision tree $\Gamma$ work. During a big step we find values of at most $M(T)$ attributes. As a result we either find the decision attached to the considered row, or localize this row in a subtable which has at most one-half of rows of the initial table.

Let during the work with the row $r$ the decision tree $\Gamma$ make $p$ big steps. After the big step number $p - 1$ the considered row will be localized in a subtable $T''$ of the table $T$. Since we must make additional big step, $N(T'') \geq 2$. It is clear that $N(T'') \leq N(T)/2^{p-1}$. Therefore, $2^p \leq N(T)$ and $p \leq \log_2 N(T)$. Taking into account that during each big step we compute values of at most $M(T)$ attributes, we conclude that the depth of $\Gamma$ is at most $M(T) \log_2 N(T)$. □

*Example 3.18.* Let us apply the considered in the proof of Theorem 3.17 procedure to the decision table $T$ depicted in Fig. 3.1. As a result we obtain the decision tree depicted in Fig. 3.2. So, $h(T) \leq 2$. When we apply the bound from this theorem to the table $T$ we obtain $h(T) \leq 2 \log_2 5$.

In the general case, it is impossible to use procedure from the proof of Theorem 3.17 as an effective algorithm for decision tree construction. When we choose for the tuple $(\sigma_1, \ldots, \sigma_n)$ attributes $f_{i_1}, \ldots, f_{i_m}$ such that $m \leq M(T)$ and $T(f_{i_1}, \sigma_{i_1}) \ldots (f_{i_m}, \sigma_{i_m})$ is a degenerate table, we (in the general case) solve an $NP$-complete problem.

It is possible to improve the bound from Theorem 3.17 and show that

$$h(T) \leq \begin{cases} M(T), & \text{if } M(T) \leq 1\,, \\ 2\log_2 N(T) + M(T), & \text{if } 2 \leq M(T) \leq 3\,, \\ \frac{M(T)\log_2 N(T)}{\log_2 M(T)} + M(T), & \text{if } M(T) \geq 4\,. \end{cases}$$

It is possible to show also that this bound does not allow for essential improvement. Corresponding results can be found in [37].

A nonempty decision table $T$ will be called a *diagnostic* table if rows of this table are labeled with pairwise different decisions. Note, that for a diagnostic table $T$ the equality $D(T) = N(T)$ holds.

**Corollary 3.19.** *Let $T$ be a diagnostic decision table. Then*

$$\max\{M(T), \log_2 N(T)\} \leq h(T) \leq M(T) \log_2 N(T) .$$

## 3.3 Conclusions

The chapter is devoted to the study of lower and upper bounds on the depth of decision trees, length of decision rules, and cardinality of tests. Bounds $h(T) \geq M(T)$ (Theorem 3.6) and $h(T) \leq M(T) \log_2 N(T)$ (Theorem 3.17) were published in [37]. Note that the parameter $M(T)$ is close to the notion of *extended teaching dimension* [24, 25], and the parameter $L(T)$ (see Theorem 3.11) is close to the notion of *teaching dimension* [22] . The algorithm of decision tree construction considered in the proof of Theorem 3.17 is in some sense similar to the "halving" algorithm [31]. The rest of results (with the exception of Theorems 3.11 and 3.14 which should be considered as "folklore") was published in [53].

We can consider not only binary decision tables filled by numbers from $\{0, 1\}$ but also $k$-valued tables filled by numbers from the set $\{0, 1, \ldots, k-1\}$ where $k > 2$. For such tables, all results considered above are true with the exception of Theorem 3.1, Theorem 3.2, and Corollary 3.19.

Instead of the bounds $h(T) \geq \log_2 D(T)$ and $h(T) \geq \log_2(R(T) + 1)$, for $k$-valued tables we will have the bounds $h(T) \geq \log_k D(T)$ and $h(T) \geq \log_k((k-1)R(T) + 1)$.

Instead of the bounds $\max\{M(T), \log_2 N(T)\} \leq h(T) \leq M(T) \log_2 N(T)$, for $k$-valued diagnostic tables we will have the bounds

$$\max\{M(T), \log_k N(T)\} \leq h(T) \leq M(T) \log_2 N(T) .$$

# 4

# Algorithms for Construction of Tests, Decision Rules and Trees

This chapter is devoted to the study of algorithms for construction of tests, decision rules and trees. Our aim is to construct tests with minimum cardinality, decision rules with minimum length, and decision trees with minimum depth. Unfortunately, all the three optimization problems are $NP$-hard. So we consider not only exact but also approximate algorithms for optimization.

The chapter consists of four sections. In Sect. 4.1, we study approximate (greedy) algorithms for optimization of tests and decision rules. These algorithms are based on greedy algorithm for the set cover problem.

Section 4.2 deals with greedy algorithm for decision tree construction.

In Sect. 4.3, we study exact algorithms for optimization of decision trees and rules which are based on dynamic programming approach. We show that if $P \neq NP$ then there is no similar algorithms for test optimization.

Section 4.4 contains conclusions.

## 4.1 Approximate Algorithms for Optimization of Tests and Decision Rules

In this section, we consider three problems of optimization connected with decision tables. *Problem of minimization of test cardinality*: for a given decision table $T$ we should construct a test for this table, which has minimum cardinality. *Problem of minimization of decision rule length*: for a given decision table $T$ and row $r$ of $T$ we need to construct a decision rule over $T$ which is true for $T$, realizable for $r$, and has minimum length. *Problem of optimization of decision rule system*: for a given decision table $T$ we should construct a complete decision rule system $S$ for $T$ with minimum value of parameter $L(S)$.

We will show that these problems are $NP$-hard and consider some results on precision of approximate polynomial algorithms for the problem solving. Also we will study greedy (approximate) algorithms for these problems. First, we consider well known set cover problem.

M. Moshkov and B. Zielosko: Combinatorial Machine Learning, SCI 360, pp. 47–67.
springerlink.com © Springer-Verlag Berlin Heidelberg 2011

### 4.1.1 Set Cover Problem

Let $A$ be a set containing $N > 0$ elements, and $F = \{S_1, \ldots, S_p\}$ be a family of subsets of the set $A$ such that $A = \bigcup_{i=1}^{p} S_i$. A subfamily $\{S_{i_1}, \ldots, S_{i_t}\}$ of the family $F$ will be called a *cover* if $\bigcup_{j=1}^{t} S_{i_j} = A$. The problem of searching for a cover with minimum cardinality $t$ is called the *set cover problem*. It is well known that this problem is an $NP$-hard problem. U. Feige [19] proved that if $NP \not\subseteq DTIME(n^{O(\log \log n)})$ then for any $\varepsilon$, $0 < \varepsilon < 1$, there is no polynomial algorithm that constructs a cover which cardinality is at most $(1-\varepsilon)C_{\min} \ln N$ where $C_{\min}$ is the minimum cardinality of a cover.

We now consider well known *greedy* algorithm for set cover problem.

Set $B := A$, and $COVER := \emptyset$.
(*) In the family $F$ we find a set $S_i$ with minimum index $i$ such that

$$|S_i \cap B| = \max\{|S_j \cap B| : S_j \in F\} .$$

Then we set $B := B \setminus S_i$ and $COVER := COVER \cup \{S_i\}$. If $B = \emptyset$ then we finish the work of the algorithm. The set $COVER$ is the result of the algorithm work. If $B \neq \emptyset$ then we return to the label (*).

We denote by $C_{\text{greedy}}$ the cardinality of the cover constructed by greedy algorithm. Remind that $C_{\min}$ is the minimum cardinality of a cover.

**Theorem 4.1.** $C_{\text{greedy}} \leq C_{\min} \ln N + 1$.

*Proof.* Denote $m = C_{\min}$. If $m = 1$ then, as it is not difficult to show, $C_{\text{greedy}} = 1$, and the considered inequality holds. Let $m \geq 2$. Let $S_i$ be a subset of maximum cardinality in the family $F$. It is clear that $|S_i| \geq N/m$ (in the opposite case $C_{\min} > m$ which is impossible). So, after the first step we will have at most $N - N/m = N(1 - 1/m)$ uncovered elements in the set $A$. After the first step we will have the following set cover problem: the set $A \setminus S_i$ and the family $\{S_1 \setminus S_i, \ldots, S_p \setminus S_i\}$. For this problem, the minimum cardinality of a cover is at most $m$. So, after the second step, when we choose a set $S_j \setminus S_i$ with maximum cardinality, the number of uncovered elements in the set $A$ will be at most $N(1 - 1/m)^2$, etc.

Let the greedy algorithm in the process of cover construction make $g$ steps (and construct a cover of cardinality $g$). Then after the step number $g-1$ we have at least one uncovered element in the set $A$. Therefore $N(1 - 1/m)^{g-1} \geq 1$ and $N \geq (1 + 1/(m-1))^{g-1}$. If we take the natural logarithm of both sides of this inequality we obtain $\ln N \geq (g-1)\ln(1 + 1/(m-1))$.

It is known that for any natural $r$ the inequality $\ln(1 + 1/r) > 1/(r+1)$ holds. Therefore $\ln N > (g-1)/m$ and $g < m \ln N + 1$. Taking into account that $m = C_{\min}$ and $g = C_{\text{greedy}}$ we obtain $C_{\text{greedy}} < C_{\min} \ln N + 1$. □

The considered bound was obtained independently by different authors: by R. G Nigmatullin [68], D. S. Johnson [26], etc. It was improved by P. Slavík in [81, 82]:

$$C_{\text{greedy}} \le C_{\min}(\ln N - \ln\ln N + 1)\ .$$

Also P. Slavík has shown that it is impossible to improve this bound essentially.

Using the mentioned result of U. Feige we obtain that if $NP \not\subseteq DTIME(n^{O(\log\log n)})$ then the greedy algorithm is close to the best (from the point of view of precision) approximate polynomial algorithms for solving the set cover problem. Unfortunately, the approximation ratio of greedy algorithm grows almost as $\ln N$.

*Example 4.2.* Consider a set cover problem depicted in Fig. 4.1. The set $A$

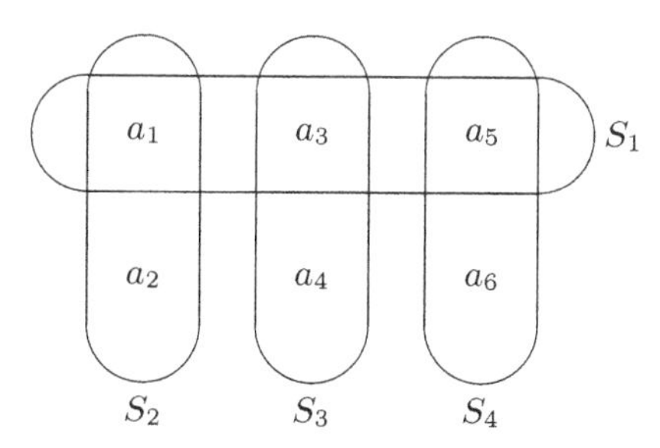


**Fig. 4.1**

consists of six elements $a_1, \ldots, a_6$, and $F = \{S_1, S_2, S_3, S_4\}$. Let us apply to this problem the greedy algorithm considered above. It is clear that this algorithm constructs the cover $\{S_1, S_2, S_3, S_4\}$. One can show that the minimum cover (cover with minimum cardinality) is $\{S_2, S_3, S_4\}$. So, sometimes the greedy algorithm constructs covers which are not optimal.

All the result relating to the set cover problem are true if we consider the class of all individual set cover problems. But often we deal with subclasses of this class. Let us consider, for example, the vertex cover problem. Let $G$ be an undirected graph. A subset $B$ of vertices of $G$ is called a *vertex cover* if for any edge of $G$ at least one vertex from $B$ is incident to this edge (an edge is incident to two vertices which are the ends of this edge). We will say that a vertex *covers* edges that are incident to this vertex.

The problem of searching for a vertex cover with minimum cardinality is known as the *vertex cover* problem. It is an $NP$-hard problem. It is known that the approximation ratio of greedy algorithm for this problem grows almost as natural logarithm on the number of edges.

Let us consider the following simple algorithm for the vertex cover problem solving. During each step we choose an uncovered edge and add the ends of this edge (two vertices which are incident to this edge) to the constructed vertex cover. It is clear that at least one end of the chosen edge must be in any vertex cover. So, the cardinality of constructed vertex cover is at most two times to the minimum cardinality of vertex cover.

### *4.1.2 Tests: From Decision Table to Set Cover Problem*

We can use the greedy algorithm for set cover problem to construct a test for given decision table $T$.

Let $T$ be a decision table containing $n$ columns labeled with $f_1, \ldots, f_n$. We consider a set cover problem $A(T), F(T) = \{S_1, \ldots, S_n\}$ where $A(T)$ is the set of all unordered pairs of rows of the table $T$ with different decisions. For $i = 1, \ldots, n$, the set $S_i$ coincides with the set of all pairs of rows from $A(T)$ which are different in the column $f_i$. One can show that the set of columns $\{f_{i_1}, \ldots, f_{i_m}\}$ is a test for the table $T$ iff the subfamily $\{S_{i_1}, \ldots, S_{i_m}\}$ is a cover for the set cover problem $A(T)$, $F(T)$.

We denote by $P(T)$ the number of unordered pairs of rows of $T$ which have different decisions. It is clear that $|A(T)| = P(T)$. It is clear also that for the considered set cover problem $C_{\min} = R(T)$.

Let us apply the greedy algorithm to the considered set cover problem. This algorithm constructs a cover which corresponds to a test for the table $T$. From Theorem 4.1 it follows that the cardinality of this test is at most

$$R(T) \ln P(T) + 1 .$$

We denote by $R_{\text{greedy}}(T)$ the cardinality of the test constructed by the following algorithm: for a given decision table $T$ we construct the set cover problem $A(T), F(T)$ and then apply to this problem the greedy algorithm for set cover problem. According to what has been said we have the following statement.

**Theorem 4.3.** *Let $T$ be a nondegenerate decision table. Then*

$$R_{\text{greedy}}(T) \le R(T) \ln P(T) + 1 .$$

It is clear that the considered algorithm for test construction has polynomial time complexity.

*Example 4.4.* Let us apply the considered algorithm to the table $T$ depicted in Fig. 3.1. For this table $A(T) = \{(1,2), (1,3), (1,4), (1,5), (2,4), (2,5), (3,4), (3,5)\}$ (we are writing here pairs of numbers of rows instead of pairs of rows), $F(T) = \{S_1, S_2, S_3\}$, $S_1 = \{(1,2), (1,4), (2,5), (3,4)\}$, $S_2 = \{(1,4), (1,5), (2,4), (2,5), (3,4), (3,5)\}$ and $S_3 = \{(1,2), (1,3), (1,5), (2,4), (3,4)\}$. At the first step, the greedy algorithm chooses $S_2$, and at the second step this algorithm chooses $S_3$. The constructed cover is $\{S_2, S_3\}$. The corresponding test is equal to $\{f_2, f_3\}$. As we know, this is the reduct with minimum cardinality.

### *4.1.3 Decision Rules: From Decision Table to Set Cover Problem*

We can apply the greedy algorithm for set cover problem to construct decision rules.

Let $T$ be a nondegenerate decision table containing $n$ columns labeled with attributes $f_1, \ldots, f_n$, and $r = (b_1, \ldots, b_n)$ be a row of $T$ labeled with a decision $t$. We consider a set cover problem $A(T, r)$, $F(T, r) = \{S_1, \ldots, S_n\}$ where $A(T, r)$ is the set of all rows of $T$ with decisions different from $t$. For $i = 1, \ldots, n$, the set $S_i$ coincides with the set of all rows from $A(T, r)$ which are different from $r$ in the column $f_i$. One can show that the decision rule

$$f_{i_1} = b_{i_1} \wedge \ldots \wedge f_{i_m} = b_{i_m} \rightarrow t$$

is true for $T$ (it is clear that this rule is realizable for $r$) if and only if the subfamily $\{S_{i_1}, \ldots, S_{i_m}\}$ is a cover for the set cover problem $A(T, r)$, $F(T, r)$.

We denote by $P(T, r)$ the number of rows of $T$ with decisions different from $t$. It is clear that $|A(T, r)| = P(T, r)$. It is clear also that for the considered set cover problem $C_{\min} = L(T, r)$.

Let us apply the greedy algorithm to the considered set cover problem. This algorithm constructs a cover which corresponds to a decision rule that is true for $T$ and realizable for $r$. From Theorem 4.1 it follows that the length of this decision rule is at most

$$L(T, r) \ln P(T, r) + 1 .$$

We denote by $L_{\text{greedy}}(T, r)$ the length of the rule constructed by the following polynomial algorithm: for a given decision table $T$ and row $r$ of $T$ we construct the set cover problem $A(T, r)$, $F(T, r)$ and then apply to this problem greedy algorithm for set cover problem. According to what has been said above we have the following statement.

**Theorem 4.5.** *Let $T$ be a nondegenerate decision table and $r$ be a row of $T$. Then*

$$L_{\text{greedy}}(T, r) \leq L(T, r) \ln P(T, r) + 1 .$$

*Example 4.6.* Let us apply the considered algorithm to the table $T$ depicted in Fig. 3.1 and to the first row of this table. For $i = 1, \ldots, 5$, we denote by $r_i$ the $i$-th row of $T$. We have $A(T, r_1) = \{r_2, r_3, r_4, r_5\}$, $F(T, r_1) = \{S_1, S_2, S_3\}$, $S_1 = \{r_2, r_4\}$, $S_2 = \{r_4, r_5\}$, $S_3 = \{r_2, r_3, r_5\}$. At the first step, the greedy algorithm chooses $S_3$, and at the second step this algorithm chooses $S_1$. The constructed cover is $\{S_1, S_3\}$. The corresponding decision rule

$$f_1 = 1 \wedge f_3 = 1 \rightarrow 1$$

has minimum length among rules over $T$ that are true for $T$ and realizable for $r_1$.

We can use the considered algorithm to construct a complete decision rule system for $T$. To this end we apply this algorithm sequentially to the table $T$ and to each row $r$ of $T$. As a result we obtain a system of rules $S$ in which each rule is true for $T$ and for every row of $T$ there exists a rule from $S$ which is realizable for this row.

We denote $L_{\text{greedy}}(T) = L(S)$ and $K(T) = \max\{P(T,r) : r \in \Delta(T)\}$, where $\Delta(T)$ is the set of rows of $T$. It is clear that $L(T) = \max\{L(T,r) : r \in \Delta(T)\}$. Using Theorem 4.5 we obtain

**Theorem 4.7.** *Let $T$ be a nondegenerate decision table. Then*

$$L_{\text{greedy}}(T) \leq L(T) \ln K(T) + 1 .$$

*Example 4.8.* Let us apply the considered algorithm to the table $T$ depicted in Fig. 3.1. As a result we obtain the following complete decision rules system for $T$:

$$S = \{f_1 = 1 \wedge f_3 = 1 \to 1, f_1 = 0 \wedge f_2 = 1 \to 2,$$
$$f_2 = 1 \wedge f_3 = 0 \to 2, f_2 = 0 \to 3, f_2 = 0 \to 3\} .$$

For this system $L(S) = 2$. We know (see example 3.13) that $L(T) = 2$.

### *4.1.4 From Set Cover Problem to Decision Table*

To understand the complexity of the problem of minimization of test cardinality we consider a reduction of an arbitrary set cover problem to the problem of minimization of test cardinality for a decision table.

Let us consider a set cover problem $A, F$ where $A = \{a_1, \ldots, a_N\}$ and $F = \{S_1, \ldots, S_m\}$. Define a decision table $T(A, F)$. This table has $m$ columns, corresponding to sets $S_1, \ldots, S_m$ respectively (these columns are labeled with $f_1, \ldots, f_m$), and $N + 1$ rows. For $j = 1, \ldots, N$, the $j$-th row corresponds to the element $a_j$. The last $(N + 1)$-th row is filled by 0. For $j = 1, \ldots, N$ and $i = 1, \ldots, m$, at the intersection of $j$-th row and $i$-th column 1 stays if and only if $a_j \in S_i$. The decision, corresponding to the last row, is equal to 2. All other rows are labeled with the decision 1.

*Example 4.9.* For the set cover problem $A, F$ considered in Example 4.2 we construct the decision table $T(A, F)$. We have $A = \{a_1, a_2, a_3, a_4, a_5, a_6\}$, $F = \{S_1, S_2, S_3, S_4\}$, $S_1 = \{a_1, a_3, a_5\}$, $S_2 = \{a_1, a_2\}$, $S_3 = \{a_3, a_4\}$ and $S_4 = \{a_5, a_6\}$. The corresponding decision table $T(A, F)$ is depicted in Fig. 4.2.

One can show that a subfamily $\{S_{i_1}, \ldots, S_{i_t}\}$ is a cover for $A, F$ if and only if the set of columns $\{f_{i_1}, \ldots, f_{i_t}\}$ is a test for the table $T(A, F)$.

So we have a polynomial time reduction of the set cover problem to the problem of minimization of test cardinality. Since the set cover problem is $NP$-hard, we have the following statement:

**Proposition 4.10.** *The problem of minimization of test cardinality is $NP$-hard.*

| | $f_1$ | $f_2$ | $f_3$ | $f_4$ | |
|---|---|---|---|---|---|
| $a_1$ | 1 | 1 | 0 | 0 | 1 |
| $a_2$ | 0 | 1 | 0 | 0 | 1 |
| $a_3$ | 1 | 0 | 1 | 0 | 1 |
| $a_4$ | 0 | 0 | 1 | 0 | 1 |
| $a_5$ | 1 | 0 | 0 | 1 | 1 |
| $a_6$ | 0 | 0 | 0 | 1 | 1 |
| | 0 | 0 | 0 | 0 | 2 |

**Fig. 4.2**

Assume that for some $\varepsilon$, $0 < \varepsilon < 1$, there exists a polynomial algorithm which for a given nondegenerate decision table $T$ constructs a test for $T$ which cardinality is at most

$$(1 - \varepsilon)R(T) \ln P(T) .$$

Let us apply this algorithm to the decision table $T(A, F)$. As a result we obtain a cover for $A, F$ which cardinality is at most $(1 - \varepsilon)C_{\min} \ln N$ (since $C_{\min} = R(T(A, F))$ and $P(T(A, F)) = N = |A|$) which is impossible if $NP \not\subseteq DTIME(n^{O(\log \log n)})$. Thus, we have

**Theorem 4.11.** *If $NP \not\subseteq DTIME(n^{O(\log \log n)})$ then for any $\varepsilon$, $0 < \varepsilon < 1$, there is no polynomial algorithm that for a given nondegenerate decision table $T$ constructs a test for $T$ which cardinality is at most*

$$(1 - \varepsilon)R(T) \ln P(T) .$$

Now we evaluate the complexity of problem of minimization of decision rule length and the complexity of problem of optimization of decision rule system.

One can show that a subfamily $\{S_{i_1}, \ldots, S_{i_t}\}$ is a cover for $A, F$ if and only if the decision rule

$$f_{i_1} = 0 \wedge \ldots \wedge f_{i_t} = 0 \rightarrow 2$$

is true for $T(A, F)$ and is realizable for the last row of the table $T(A, F)$. So we have a polynomial time reduction of the set cover problem to the problem of minimization of decision rule length. Since the set cover problem is $NP$-hard, we have the following statement:

**Proposition 4.12.** *The problem of minimization of decision rule length is $NP$-hard.*

Let us assume that for some $\varepsilon$, $0 < \varepsilon < 1$, there exists a polynomial algorithm which for a given nondegenerate decision table $T$ and row $r$ of $T$ constructs a decision rule which is true for $T$ and realizable for $r$, and which length is at most

$$(1-\varepsilon)L(T,r)\ln P(T,r)\ .$$

Let us apply this algorithm to the decision table $T(A,F)$ and the last row $r$ of $T(A,F)$. As a result we obtain a cover for $A,F$ which cardinality is at most $(1-\varepsilon)C_{\min}\ln N$ (since $C_{\min}=L(T(A,F),r)$ and $P(T(A,F),r)=N=|A|$) which is impossible if $NP \not\subseteq DTIME(n^{O(\log\log n)})$. Thus, we have

**Theorem 4.13.** *If $NP \not\subseteq DTIME(n^{O(\log\log n)})$ then for any $\varepsilon$, $0<\varepsilon<1$, there is no polynomial algorithm that for a given nondegenerate decision table $T$ and row $r$ of $T$ constructs a decision rule which is true for $T$, realizable for $r$, and which length is at most*

$$(1-\varepsilon)L(T,r)\ln P(T,r)\ .$$

Let us consider the decision table $T(A,F)$. For $j=1,\ldots,N+1$, we denote by $r_j$ the $j$-th row of $T(A,F)$. Let $j\in\{1,\ldots,N\}$. We know that there exists a subset $S_i\in F$ such that $a_j\in S_i$. Therefore the decision rule

$$f_i=1\rightarrow 1$$

is true for $T(A,F)$ and realizable for $r_j$. Thus $L(T(A,F),r_j)=1$ for any $j\in\{1,\ldots,N\}$. Hence $L(T(A,F))=L(T(A,F),r)$ where $r=r_{N+1}$. So if we find a complete decision rule system $S$ for $T(A,F)$ such that $L(S)=L(T(A,F))$ then in this system we will find a decision rule of the kind

$$f_{i_1}=0\wedge\ldots\wedge f_{i_t}=0\rightarrow 2$$

for which $t=L(T(A,F),r)$. We know that $\{S_{i_1},\ldots,S_{i_t}\}$ is a set cover for $A,F$ with minimum cardinality. Since the set cover problem is $NP$-hard, we have the following statement:

**Proposition 4.14.** *The problem of optimization of decision rule system is $NP$-hard.*

Let us assume that for some $\varepsilon$, $0<\varepsilon<1$, there exists a polynomial algorithm which for a given nondegenerate decision table $T$ constructs a complete decision rule system $S$ for $T$ such that

$$L(S)\leq(1-\varepsilon)L(T)\ln K(T)\ .$$

Let us apply this algorithm to the decision table $T(A,F)$. In the constructed complete decision rule system for $T(A,F)$ we will find a decision rule of the kind

$$f_{i_1}=0\wedge\ldots\wedge f_{i_t}=0\rightarrow 2$$

which is true for $T$ and realizable for the last row $r$ of the table $T(A,F)$. We know that $\{S_{i_1},\ldots,S_{i_t}\}$ is a cover for $A,F$ and $t\leq(1-\varepsilon)L(T(A,F))\ln K(T)$. We know also that $L(T(A,F))=L(T(A,F),r)=C_{\min}$ and $K(T)=N$. Therefore we have a polynomial algorithm that for a given set cover problem

$A, F$ constructs a cover for $A, F$ which cardinality is at most $(1-\varepsilon)C_{\min} \ln N$, which is impossible if $NP \not\subseteq DTIME(n^{O(\log\log n)})$.

Thus, we have

**Theorem 4.15.** *If $NP \not\subseteq DTIME(n^{O(\log\log n)})$ then for any $\varepsilon$, $0 < \varepsilon < 1$, there is no polynomial algorithm that for a given nondegenerate decision table $T$ constructs a complete decision rule system $S$ for $T$ such that*

$$L(S) \leq (1-\varepsilon)L(T)\ln K(T) .$$

## 4.2 Approximate Algorithm for Decision Tree Optimization

In this section, we study *problem of minimization of decision tree depth*: for a given decision table $T$ it is required to construct a decision tree for this table which has minimum depth.

This problem is $NP$-hard. We will consider some results on precision of polynomial approximate algorithms for the problem solving and will concentrate on the study of greedy algorithm for decision tree depth minimization.

We now describe an algorithm $U$ which for a decision table $T$ constructs a decision tree $U(T)$ for the table $T$. Let $T$ have $n$ columns labeled with attributes $f_1, \ldots, f_n$.

*Step* 1: Construct a tree consisting of a single node labeled with the table $T$ and proceed to the second step.

Suppose $t \geq 1$ steps have been made already. The tree obtained at the step $t$ will be denoted by $G$.

*Step* $(t+1)$: If no one node of the tree $G$ is labeled with a table then we denote by $U(T)$ the tree $G$. The work of the algorithm $U$ is completed.

Otherwise, we choose certain node $v$ in the tree $G$ which is labeled with a subtable of the table $T$. Let the node $v$ be labeled with the table $T'$. If $T'$ is a degenerate table (all rows of the table are labeled with the same decision $d$) then instead of $T'$ we mark the node $v$ by the number $d$ and proceed to the step $(t+2)$. Let $T'$ be a nondegenerate table. Then, for $i = 1, \ldots, n$, we compute the value

$$Q(f_i) = \max\{P(T'(f_i, 0)), P(T'(f_i, 1))\} .$$

We mark the node $v$ by the attribute $f_{i_0}$ where $i_0$ is the minimum $i$ for which $Q(f_i)$ has minimum value. For each $\delta \in \{0, 1\}$, we add to the tree $G$ the node $v(\delta)$, mark this node by the table $T'(f_{i_0}, \delta)$, draw the edge from $v$ to $v(\delta)$, and mark this edge by $\delta$. Proceed to the step $(t+2)$.

*Example 4.16.* Let us apply the algorithm $U$ to the decision table $T$ depicted in Fig. 3.1. After the first step, we obtain the tree which has only one node $v$ that is labeled with the table $T$. The table $T$ is not degenerate. So, for $i = 1, 2, 3$, we compute the value

$$Q(f_i) = \max\{P(T(f_i, 0)), P(T(f_i, 1))\} .$$

It is not difficult to see that $Q(f_1) = \max\{1, 3\} = 3$, $Q(f_2) = \max\{0, 2\} = 2$, and $Q(f_3) = \max\{2, 1\} = 2$. It is clear that 2 is the minimum index for which the value of $Q(f_2)$ is minimum. So, after the second step we will have the tree $G$ depicted in Fig. 4.3. We omit next steps. One can show that as a result of the algorithm $U$ work for the table $T$, we will obtain the tree $U(T)$ depicted in Fig. 3.2.

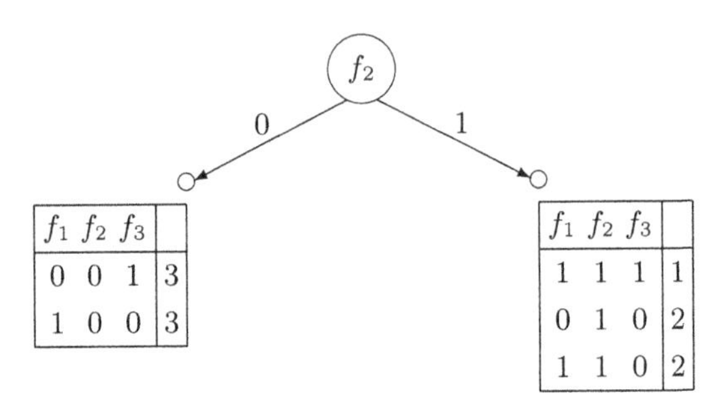


**Fig. 4.3**

Now, we evaluate the number of steps which the algorithm $U$ makes during the construction of the decision tree $U(T)$.

**Theorem 4.17.** *Let $T$ be a decision table. Then during the construction of the tree $U(T)$ the algorithm $U$ makes at most $2N(T) + 1$ steps.*

*Proof.* One can show that for each terminal node of the tree $U(T)$ there exists a row of $T$ for which the work of $U(T)$ finishes in this node. Therefore the number of terminal nodes in $U(T)$ is at most $N(T)$. It is not difficult to prove that the number of working nodes in $U(T)$ is equal to the number of terminal nodes minus 1. Simple analysis of the algorithm $U$ work shows that the number of steps of $U$ in the process of the tree $U(T)$ construction is equal to the number of nodes in $U(T)$ plus 2. Therefore the number of steps is bounded from above by $2N(T) + 1$. □

Using this theorem it is not difficult to prove that the algorithm $U$ has polynomial time complexity.

We now consider a lemma which will be used later in the proof of a bound on algorithm $U$ precision (as an algorithm for minimization of decision tree depth).

**Lemma 4.18.** *Let $T$ be a decision table, $T'$ be a subtable of $T$, $f_i$ be an attribute attached to a column of $T$, and $\delta \in \{0, 1\}$. Then*

$$P(T) - P(T(f_i, \delta)) \geq P(T') - P(T'(f_i, \delta)) .$$

*Proof.* Denote by $B$ (respectively by $B'$) the set of pairs of rows of $T$ (respectively of $T'$) which have different decisions and in each of which at least one row has in the column $f_i$ the number that is not equal to $\delta$. One can show that $B' \subseteq B$, $|B'| = P(T') - P(T'(f_i, \delta))$ and $|B| = P(T) - P(T(f_i, \delta))$. □

**Theorem 4.19.** *Let $T$ be a nondegenerate decision table. Then*

$$h(U(T)) \leq M(T) \ln P(T) + 1 .$$

*Proof.* Let $T$ be a table with $n$ columns labeled with attributes $f_1, \ldots, f_n$. For $i = 1, \ldots, n$, denote by $\sigma_i$ a number from $\{0, 1\}$ such that $P(T(f_i, \sigma_i)) = \max\{P(T(f_i, \sigma)) : \sigma \in \{0, 1\}\}$. It is clear that the root of the tree $U(T)$ is labeled with attribute $f_{i_0}$ where $i_0$ is the minimum $i$ for which $P(T(f_i, \sigma_i))$ has minimum value. Of course, $Q(f_i) = P(T(f_i, \sigma_i))$.

Let us show that

$$P(T(f_{i_0}, \sigma_{i_0})) \leq (1 - 1/M(T))\, P(T) .$$

It is clear that there exist attributes $f_{i_1}, \ldots, f_{i_m} \in \{f_1, \ldots, f_n\}$ such that $T(f_{i_1}, \sigma_{i_1}) \ldots (f_{i_m}, \sigma_{i_m})$ is a degenerate table and $m \leq M(T)$. Evidently, $P(T(f_{i_1}, \sigma_{i_1}) \ldots (f_{i_m}, \sigma_{i_m})) = 0$. Therefore $P(T) - [P(T) - P(T(f_{i_1}, \sigma_{i_1}))] - [P(T(f_{i_1}, \sigma_{i_1})) - P(T(f_{i_1}, \sigma_{i_1})(f_{i_2}, \sigma_{i_2}))] - \ldots - [P(T(f_{i_1}, \sigma_{i_1}) \ldots (f_{i_{m-1}}, \sigma_{i_{m-1}})) - P(T(f_{i_1}, \sigma_{i_1}) \ldots (f_{i_m}, \sigma_{i_m}))] = P(T(f_{i_1}, \sigma_{i_1}) \ldots (f_{i_m}, \sigma_{i_m})) = 0$.

From Lemma 4.18 it follows that, for $j = 1, \ldots, m-1$, $P(T(f_{i_1}, \sigma_{i_1}) \ldots (f_{i_j}, \sigma_{i_j})) - P(T(f_{i_1}, \sigma_{i_1}) \ldots (f_{i_j}, \sigma_{i_j})(f_{i_{j+1}}, \sigma_{i_{j+1}})) \leq P(T) - P(T(f_{i_{j+1}}, \sigma_{i_{j+1}}))$.

Therefore $P(T) - \sum_{j=1}^{m} (P(T) - P(T(f_{i_j}, \sigma_{i_j}))) \leq 0$. Since $P(T(f_{i_0}, \sigma_{i_0})) \leq P(T(f_{i_j}, \sigma_{i_j}))$, $j = 1, \ldots, m$, we have $P(T) - m(P(T) - P(T(f_{i_0}, \sigma_{i_0}))) \leq 0$ and $P(T(f_{i_0}, \sigma_{i_0})) \leq (1 - 1/m)P(T)$. Taking into account that $m \leq M(T)$ we obtain $P(T(f_{i_0}, \sigma_{i_0})) \leq (1 - 1/M(T))\, P(T)$.

Assume that $M(T) = 1$. From the obtained inequality and from the description of the algorithm $U$ it follows that $h(U(T)) = 1$. So, if $M(T) = 1$ then the statement of theorem is true.

Let now $M(T) \geq 2$. Consider a longest path in the tree $U(T)$ from the root to a terminal node. Let its length be equal to $k$. Let working nodes of this path be labeled with attributes $f_{j_1}, \ldots, f_{j_k}$, where $f_{j_1} = f_{i_0}$, and edges be labeled with numbers $\delta_1, \ldots, \delta_k$. For $t = 1, \ldots, k$, we denote by $T_t$ the table $T(f_{j_1}, \delta_1) \ldots (f_{j_t}, \delta_t)$. From Lemma 3.4 it follows that $M(T_t) \leq M(T)$ for $t = 1, \ldots, k$. We have proved that $P(T_1) \leq P(T)(1 - 1/M(T))$.

One can prove in the same way that $P(T_t) \leq P(T)(1 - 1/M(T))^t$. Consider the table $T_{k-1}$. For this table $P(T_{k-1}) \leq P(T)(1 - 1/M(T))^{k-1}$. Using the description of the algorithm $U$ we conclude that $T_{k-1}$ is a nondegenerate table. Therefore $P(T_{k-1}) \geq 1$. So, we have $1 \leq P(T)(1 - 1/M(T))^{k-1}$, $(M(T)/(M(T) - 1))^{k-1} \leq P(T)$ and $(1 + 1/(M(T) - 1))^{k-1} \leq P(T)$. If we take natural logarithm of both sides of this inequality we obtain $(k - 1) \ln(1 + 1/(M(T) - 1)) \leq \ln P(T)$. It is known that for any natural $r$ the inequality $\ln(1 + 1/r) > 1/(r + 1)$ holds. Since $M(T) \geq 2$, we obtain

$(k-1)/M(T) < \ln P(T)$ and $k < M(T)\ln P(T)+1$. Taking into account that $k = h(U(T))$ we obtain $h(U(T)) < M(T)\ln P(T) + 1$. □

Using the bound from Theorem 3.6 we obtain the following

**Corollary 4.20.** *For any nondegenerate decision table $T$*

$$h(U(T)) \le h(T)\ln P(T) + 1 .$$

It is possible to improve the considered bounds and show that for any decision table $T$

$$h(U(T)) \le \begin{cases} M(T), & \text{if } M(T) \le 1 , \\ M(T)(\ln P(T) - \ln M(T) + 1), & \text{if } M(T) \ge 2 , \end{cases}$$

and

$$h(U(T)) \le \begin{cases} h(T), & \text{if } h(T) \le 1 , \\ h(T)(\ln P(T) - \ln h(T) + 1), & \text{if } h(T) \ge 2 . \end{cases}$$

The last two bounds do not allow for essential improvement (see details in [37]).

To understand the complexity of the problem of decision tree depth minimization, we consider a reduction of an arbitrary set cover problem to the problem of minimization of decision tree depth for a decision table.

Let us consider a set cover problem $A, F$ where $A = \{a_1, \dots, a_N\}$ and $F = \{S_1, \dots, S_m\}$, and the decision table $T(A, F)$ described above. This table has $m$ columns, corresponding to sets $S_1, \dots, S_m$ respectively (these columns are labeled with $f_1, \dots, f_m$), and $N + 1$ rows. For $j = 1, \dots, N$, the $j$-th row corresponds to the element $a_j$. The last $(N + 1)$-th row is filled by 0. For $j = 1, \dots, N$ and $i = 1, \dots, m$, at the intersection of $j$-th row and $i$-th column 1 stays if and only if $a_j \in S_i$. The decision, attached to the last row, is equal to 2. All other rows are labeled with the decision 1. One can show that $P(T(A, F)) = N$ and $h(T(A, F)) = C_{\min}$ where $C_{\min}$ is the minimum cardinality of a cover for the considered set cover problem.

**Proposition 4.21.** *The problem of minimization of decision tree depth is $NP$-hard.*

*Proof.* Let $A, F$ be an arbitrary set cover problem. It is clear that in polynomial time we can construct the decision table $T(A, F)$. Let $\Gamma$ be a decision tree for $T(A, F)$ such that $h(\Gamma) = h(T(A, F))$. Consider the path in this tree from the root to a terminal node in which each edge is labeled with 0. Let $\{f_{i_1}, \dots, f_{i_t}\}$ be the set of attributes attached to working nodes of this path. One can show that $\{S_{i_1}, \dots, S_{i_t}\}$ is a cover with minimum cardinality for the problem $A, F$. So, we have a polynomial time reduction of the set cover problem to the problem of minimization of decision tree depth. Taking into account that the set cover problem is $NP$-hard we obtain that the problem of minimization of decision tree depth is $NP$-hard too. □

**Theorem 4.22.** *If $NP \not\subseteq DTIME(n^{O(\log \log n)})$ then for any $\varepsilon > 0$ there is no polynomial algorithm which for a given nondegenerate decision table $T$ constructs a decision tree for $T$ which depth is at most $(1 - \varepsilon)h(T) \ln P(T)$.*

*Proof.* Let us assume that $NP \not\subseteq DTIME(n^{O(\log \log n)})$ and such an algorithm exists. Let $A, F$ be an arbitrary set cover problem. Construct the decision table $T(A, F)$ and apply to this table the considered algorithm. As a result we obtain a decision tree $\Gamma$ for $T(A, F)$ such that $h(\Gamma) \leq (1 - \varepsilon)h(T(A, F)) \ln P(T(A, F))$. We know that $h(T(A, F)) = C_{\min}$ which is the minimum cardinality of a cover for the problem $A, F$, and $P(T(A, F)) = N$ where $N = |A|$.

Consider the path in $\Gamma$ in which all edges are labeled with 0. Let $\{f_{i_1}, \dots, f_{i_t}\}$ be the set of attributes attached to nodes of this path. Then the set $\{S_{i_1}, \dots, S_{i_t}\}$ is a cover for the problem $A, F$. The cardinality of this cover is at most $(1 - \varepsilon)C_{\min} \ln N$. But this contradicts to the results of U. Feige considered in Sect. 4.1.1. □

## 4.3 Exact Algorithms for Optimization of Trees, Rules and Tests

In this subsection, we consider some possibilities (mainly based on dynamic programming) to design exact algorithms for optimization of trees, rules and tests.

### *4.3.1 Optimization of Decision Trees*

We consider an algorithm based on dynamic programming approach which for a given decision table constructs a decision tree for this table with minimum depth (see also [55]). Of course, in the worst case the considered algorithm has exponential time complexity. However, later (when we consider local approach to decision tree investigation over infinite information systems) we will describe infinite information systems for each of which this algorithm has (for decision tables over the considered system) polynomial time complexity depending on the number of columns (attributes) in the table.

The idea of algorithm is very simple. Let $T$ be a decision table with $n$ columns labeled with attributes $f_1, \dots, f_n$. If $T$ is a degenerate decision table which rows are labeled with the same decision $d$, the tree, which has exactly one node labeled with $d$, is an optimal decision tree for $T$. Let $T$ be a nondegenerate table. In this case, $h(T) \geq 1$, and the root of any optimal decision tree for the table $T$ is labeled with an attribute. Denote by $E(T)$ the set of columns (attributes) of $T$ which have both 0 and 1 values. Of course, we must consider only attributes from $E(T)$.

What can we say about the minimum depth of a decision tree for the table $T$ in which the root is labeled with the attribute $f_i \in E(T)$? We denote the minimum depth of such a tree by $h(T, f_i)$. It is clear that

$$h(T, f_i) = 1 + \max\{h(T(f_i, 0)), h(T(f_i, 1))\} .$$

So, if we know values of $h(T(f_i, \delta))$ for any $f_i \in E(T)$ and $\delta \in \{0, 1\}$, we can find the value

$$h(T) = \min\{h(T, f_i) : f_i \in E(T)\}$$

and at least one attribute $f_i$ for which $h(T) = h(T, f_i)$.

How can we construct an optimal decision tree (decision tree with minimum depth), using attribute $f_i$ in the capacity of an attribute attached to the root? Let $\Gamma_0$ and $\Gamma_1$ be optimal decision trees for the subtables $T(f_i, 0)$ and $T(f_i, 1)$ respectively. Then the decision tree depicted in Fig. 4.4 is an optimal decision tree for the table $T$. So, if we know optimal decision trees

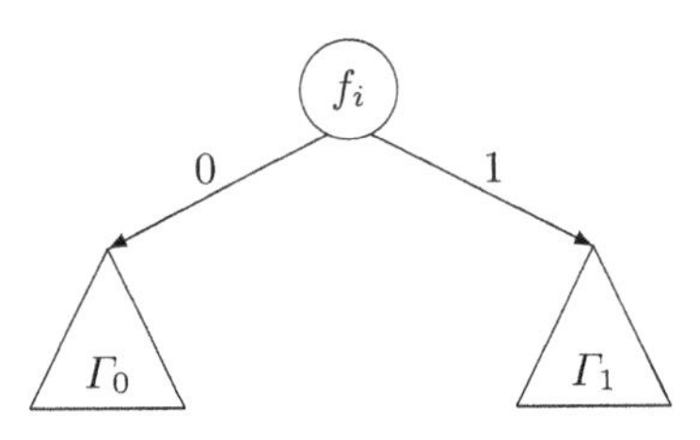


**Fig. 4.4**

for subtables of the table $T$ we can construct an optimal decision tree for the table $T$.

We now describe the algorithm for minimization of decision tree depth more formally. We denote this algorithm by $W$. A nonempty subtable $T'$ of the table $T$ will be called a *separable* subtable of $T$ if there exist attributes $f_{i_1}, \ldots, f_{i_t}$ from $\{f_1, \ldots, f_n\}$ and numbers $\delta_1, \ldots, \delta_t$ from $\{0, 1\}$ such that $T' = T(f_{i_1}, \delta_1) \ldots (f_{i_t}, \delta_t)$. We denote by $SEP(T)$ the set of all separable subtables of the table $T$ including $T$.

**The first part** of the algorithm $W$ work is the construction of the set $SEP(T)$.

*Step* 1: We set $SEP(T) = \{T\}$ and pass to the second step. After the first step $T$ is not labeled as a treated table.

Suppose $t \geq 1$ steps have been made already.

*Step* $(t+1)$: Let all tables in the set $SEP(T)$ are labeled as treated tables. In this case, we finish the first part of the algorithm $W$ work. Let there exist a table $D \in SEP(T)$ which is not treated. We add to the set $SEP(T)$ all subtables of the kind $D(f_i, \delta)$, where $f_i \in E(D)$ and $\delta \in \{0, 1\}$, which were not in $SEP(T)$, mark the table $D$ as treated and pass to the step $(t + 2)$.

It is clear that during the first part the algorithm $W$ makes exactly $|SEP(T)| + 2$ steps.

**The second part** of the algorithm $W$ work is the construction of an optimal decision tree $W(T)$ for the table $T$. Beginning with smallest subtables from $SEP(T)$, the algorithm $W$ at each step will correspond to a subtable from $SEP(T)$ an optimal decision tree for this subtable.

Suppose that $p \geq 0$ steps of the second part of algorithm $W$ have been made already.

*Step* $(p+1)$: If the table $T$ in the set $SEP(T)$ is labeled with a decision tree then this tree is the result of the algorithm $W$ work (we denote this tree by $W(T)$). Otherwise, choose in the set $SEP(T)$ a table $D$ satisfying the following conditions:

a) the table $D$ is not labeled with a decision tree;
b) either $D$ is a degenerate table, or a nondegenerate table such that all separable subtables of $D$ of the kind $D(f_i, \delta)$, $f_i \in E(D)$, $\delta \in \{0,1\}$, are labeled with decision trees.

Let $D$ be a degenerate table in which all rows are labeled with the same decision $d$. Then we mark the table $D$ by the decision tree consisting of one node which is labeled with the number $d$.

Otherwise, for any $f_i \in E(D)$ we construct a decision tree $\Gamma(f_i)$. The root of this tree is labeled with the attribute $f_i$. The root is the initial node of exactly two edges which are labeled with 0 and 1. These edges enter to roots of decision trees $\Gamma(f_i, 0)$ and $\Gamma(f_i, 1)$ respectively, where $\Gamma(f_i, 0)$ and $\Gamma(f_i, 1)$ are decision trees attached to tables $D(f_i, 0)$ and $D(f_i, 1)$. Mark the table $D$ by one of the trees $\Gamma(f_j)$, $f_j \in E(D)$, having minimum depth, and proceed to the step $(p+2)$.

It is clear that during the second part, the algorithm $W$ makes exactly $|SEP(T)+1|$ steps.

It is not difficult to prove the following statement.

**Theorem 4.23.** *For any nondegenerate decision table $T$ the algorithm $W$ constructs a decision tree $W(T)$ for the table $T$ such that $h(W(T)) = h(T)$, and makes exactly $2|SEP(T)|+3$ steps. The time of the algorithm $W$ work is bounded from below by $|SEP(T)|$, and bounded from above by a polynomial on $|SEP(T)|$ and on the number of columns in the table $T$.*

*Example 4.24.* Let us apply the algorithm $W$ to the table $T$ depicted in Fig. 3.1. As a result, we obtain the set $SEP(T)$ in which each subtable is labeled with an optimal decision tree for this subtable (see Fig. 4.5).

Similar algorithms exist also for other complexity measures, for example, for average depth of decision trees and number of nodes in decision trees [11, 10, 12].

### *4.3.2 Optimization of Decision Rules*

We now describe an algorithm $V$ for minimization of the length of decision rules (see also [97]). The algorithm $V$ work consists of two parts. The first

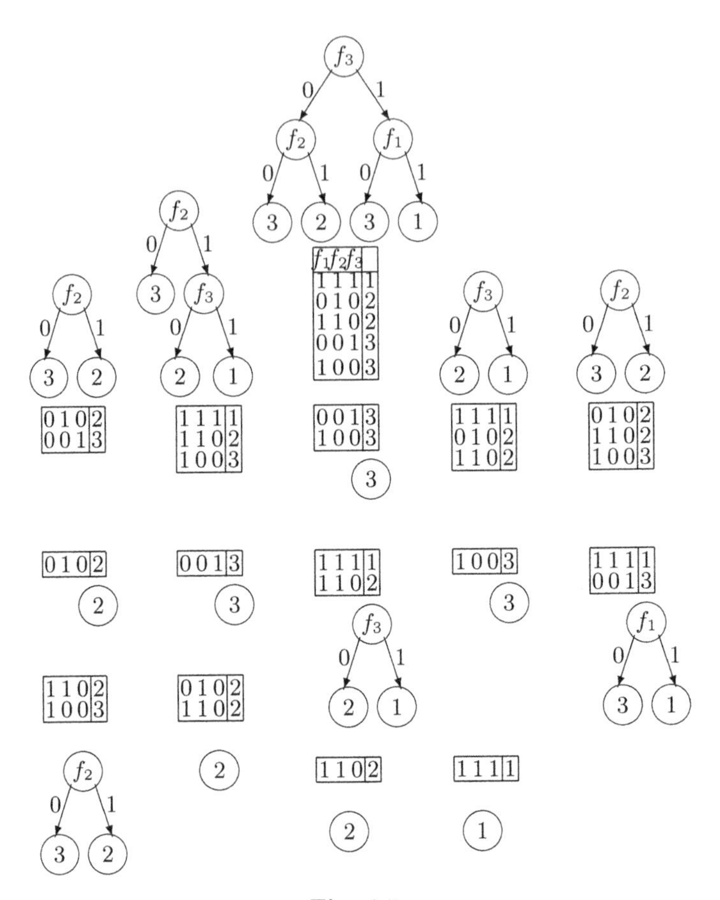

Fig. 4.5

part is the construction of the set $SEP(T)$. This part coincides with the first part of the algorithm $W$.

**The second part** of the algorithm $V$ work is the construction of an optimal decision rule $V(T, r)$ for each row $r$ of $T$. Beginning with the smallest subtables from $SEP(T)$, the algorithm $V$ at each step will correspond to each row $r$ of a subtable $T'$ an optimal decision rule for $T'$ and $r$ (an *optimal* rule for $T'$ and $r$ means a decision rule with minimum length which is true for $T'$ and realizable for $r$).

Suppose $p \geq 0$ steps of the second part of algorithm $V$ have been made already.

*Step* $(p + 1)$. If each row $r$ of the table $T$ is labeled with a decision rule, then the rule attached to $r$ is the result of the work of $V$ for $T$ and $r$ (we denote this rule by $V(T, r)$). Otherwise, choose in the set $SEP(T)$ a table $D$ satisfying the following conditions:

a) rows of $D$ are not labeled with decision rules;
b) either $D$ is a degenerate table, or a nondegenerate table such that for all separable subtables of $D$ of the kind $D(f_i, \delta)$, $f_i \in E(D)$, $\delta \in \{0,1\}$, each row is labeled with a decision rule.

Let $D$ be a degenerate table in which all rows are labeled with the same decision $d$. Then we mark each row of $D$ by the decision rule $\rightarrow d$.

Let $D$ be a nondegenerate decision table and $r = (\delta_1, \ldots, \delta_n)$ be a row of $D$ labeled with a decision $d$. For any $f_i \in E(D)$ we construct a rule rule$(r, f_i)$. Let the row $r$ in the table $D(f_i, \delta_i)$ be labeled with the rule $\alpha_i \rightarrow d$. Then the rule rule$(r, f_i)$ is equal to $f_i = \delta_i \wedge \alpha_i \rightarrow d$. We mark the row $r$ of the table $D$ with one of the rules rule$(r, f_i)$, $f_i \in E(D)$, having minimum length. We denote this rule by $V(T, r)$. We attach a rule to each row of $D$ in the same way, and proceed to the step $(p+2)$. It is clear that during the second part the algorithm $V$ makes exactly $|SEP(T)|+1$ steps.

**Lemma 4.25.** *For any table $D \in SEP(T)$ and any row $r$ of $D$ the decision rule attached to $r$ after the end of algorithm $V$ work is true for $D$, realizable for $r$ and has minimum length $L(D, r)$.*

*Proof.* We will prove this statement by induction on tables from $SEP(T)$. It is clear that for each degenerate table $D$ from $SEP(T)$ the considered statement is true.

Let $D \in SEP(T)$, $D$ be a nondegenerate table and for each $f_i \in E(D)$ and $\delta \in \{0,1\}$ the considered statement hold for the table $D(f_i, \delta)$. Let $r = (\delta_1, \ldots, \delta_n)$ be a row of $D$. Then for some $f_i \in E(D)$ the row $r$ is labeled with the decision rule $f_i = \delta_i \wedge \alpha_i \rightarrow d$ where $d$ is the decision attached to $r$ and $\alpha_i \rightarrow d$ is the decision rule attached to row $r$ in the table $D(f_i, \delta_i)$. According to the inductive hypothesis, the rule $\alpha_i \rightarrow d$ is true for $D(f_i, \delta_i)$ and is realizable for $r$. Therefore the rule $f_i = \delta_i \wedge \alpha_i \rightarrow d$ is true for $D$ and realizable for $r$.

Let us assume that there exists a shorter decision rule which is true for $D$ and realizable for $r$. Since $D$ is nondegenerate, the left-hand side of this rule should contain an equality of the kind $f_j = \delta_j$ for some $f_j \in E(D)$. Therefore this rule can be represented in the form $f_j = \delta_j \wedge \beta \rightarrow d$. Since this rule is true for $D$ and realizable for $r$, the rule $\beta \rightarrow d$ is true for $D(f_j, \delta_j)$ and realizable for $r$. According to the inductive hypothesis, the row $r$ in the table $D(f_j, \delta_j)$ is labeled with a rule $\gamma \rightarrow d$ which length is at most the length of the rule $\beta \rightarrow d$. From here and from the description of the algorithm $V$ it follows that the row $r$ in $D$ is labeled with a rule which length is at most the length of the rule $f_j = \delta_j \wedge \gamma \rightarrow d$ which is impossible. Therefore, the rule attached to the row $r$ in $D$ has minimum length among rules which are true for $D$ and realizable for $r$. □

Using this lemma it is not difficult to prove the following statement:

**Theorem 4.26.** *For any nondegenerate decision table $T$ and any row $r$ of $T$ the algorithm $V$ constructs a decision rule $V(T,r)$ which is true for $T$, realizable for $r$, and has minimum length $L(T,r)$. During the construction of optimal rules for rows of $T$, the algorithm $V$ makes exactly $2|SEP(T)|+3$ steps. The time of the algorithm $V$ work is bounded from below by $|SEP(T)|$, and bounded from above by a polynomial on $|SEP(T)|$ and on the number of columns in the table $T$.*

### 4.3.3 Optimization of Tests

It would be very well to have similar algorithm for the problem of minimization of test cardinality. Unfortunately, if $P \neq NP$, then such algorithms do not exist. To prove this result, we consider an interesting example of a class of decision tables with relatively small number of separable subtables (see also [12]).

Let we have a finite set $S = \{(a_1,b_1),\ldots,(a_n,b_n)\}$ of points in the plane and a mapping $\mu$ which colors these points into two colors: green and white. We must separate white points from green ones. To this end, we can use straight lines which are defined by equations of the kind $x = \alpha$ or $y = \beta$. It is known that the problem of construction of minimum separating set (a set of straight lines with minimum cardinality which separate green and white points) is $NP$-hard (result of B.S. Chlebus and S.H. Nguyen [15]).

Let us transform this problem into a problem of minimization of test cardinality. It is possible that $a_i = a_j$ or $b_i = b_j$ for different $i$ and $j$. Let $a_{i_1},\ldots,a_{i_m}$ be all pairwise different numbers from the set $\{a_1,\ldots,a_n\}$ which are ordered such that $a_{i_1} < \ldots < a_{i_m}$. Let $b_{j_1},\ldots,b_{j_t}$ be all pairwise different numbers from the set $\{b_1,\ldots,b_n\}$ which are ordered such that $b_{j_1} < \ldots < b_{j_t}$.

It is clear that there exists a minimum separating set which is a subset of the set of straight lines defined by equations $x = a_{i_1}-1$, $x = (a_{i_1}+a_{i_2})/2,\ldots$, $x = (a_{i_{m-1}} + a_{i_m})/2$, $x = a_{i_m} + 1$, $y = b_{j_1} - 1$, $y = (b_{j_1} + b_{j_2})/2,\ldots$, $y = (b_{j_{t-1}} + b_{j_t})/2$, $y = b_{j_t} + 1$.

Let us correspond an attribute to each such straight line. This attribute is defined on the set $S$ and takes values from the set $\{0,1\}$. Consider the line defined by equation $x = \alpha$. Then the value of corresponding attribute is equal to 0 on a point $(a,b) \in S$ if and only if $a < \alpha$. Consider the line defined by equation $y = \beta$. Then the value of corresponding attribute is equal to 0 if and only if $b < \beta$.

Let us construct a decision table $T(S,\mu)$ with $m+t+2$ columns and $n$ rows. Columns are labeled with attributes $f_1,\ldots,f_{m+t+2}$, corresponding to the considered $m+t+2$ lines. Attributes $f_1,\ldots,f_{m+1}$ correspond to lines defined by equations $x = a_{i_1} - 1$, $x = (a_{i_1} + a_{i_2})/2,\ldots,x = (a_{i_{m-1}} + a_{i_m})/2, x = a_{i_m} + 1$ respectively. Attributes $f_{m+2},\ldots,f_{m+t+2}$ correspond to lines defined by equations $y = b_{j_1} - 1$, $y = (b_{j_1} + b_{j_2})/2,\ldots,y = (b_{j_{t-1}} + b_{j_t})/2, y = b_{j_t} + 1$ respectively. Rows of the table $T(S,\mu)$ correspond to points $(a_1,b_1),\ldots,(a_n,b_n)$. At the intersection of the column $f_l$ and row

$(a_p, b_p)$ the value $f_l(a_p, b_p)$ stays. For $p = 1, \ldots, n$, the row $(a_p, b_p)$ is labeled with the decision 1 if $(a_p, b_p)$ is a white point, and with the decision 2 if $(a_p, b_p)$ is a green point.

It is clear that the minimum cardinality of a separating set is equal to the minimum cardinality of a test for the table $T(S, \mu)$. Also, it is clear that there is a polynomial algorithm which for a given set of points $S$ and given mapping $\mu$ constructs the decision table $T(S, \mu)$. Since the problem of minimization of separating set cardinality is $NP$-hard, the problem of minimization of test cardinality for decision tables of the kind $T(S, \mu)$ is $NP$-hard.

Let us evaluate the number of separable subtables of the table $T(S, \mu)$. It is not difficult to show that each separable subtable $T'$ of this table can be represented in the following form: $T' = T(f_i, 1)(f_j, 0)(f_l, 1)(f_k, 0)$, where $T = T(S, \mu)$ and $1 \leq i < j \leq m + 1 < l < k \leq t + m + 2$.

Therefore the number of separable subtables of the table $T(S, \mu)$ is at most $(m + t + 2)^4$. So, the number of separable subtables is bounded from above by a polynomial on the number of columns in the table. Note that $(m + t + 2) \leq 2n + 2$.

Now we can prove the following statement.

**Theorem 4.27.** *If $P \neq NP$, then there is no algorithm which for a given decision table $T$ constructs a test for $T$ with minimum cardinality, and for which the time of work is bounded from above by a polynomial depending on the number of columns in $T$ and the number of separable subtables of $T$.*

*Proof.* Assume that $P \neq NP$ but such an algorithm exists. Then we can construct a polynomial algorithm for $NP$-hard problem of minimization of separating set cardinality, but it is impossible. □

It should be pointed out that the problem of decision tree depth minimization can be solved for decision tables of the kind $T(S, \mu)$ by the algorithm $W$ which for such tables has polynomial time complexity depending on the number of columns in the table.

Tables $T(S, \mu)$ have interesting property which will be used in our examples.

**Proposition 4.28**

$$M(T(S, \mu)) \leq 4 .$$

*Proof.* Denote $T = T(S, \mu)$. Let $\bar{\delta} = (\delta_1, \ldots, \delta_{m+t+2}) \in \{0, 1\}^{m+t+2}$. If $\delta_1 = 0$, or $\delta_{m+1} = 1$, or $\delta_{m+2} = 0$, or $\delta_{m+t+2} = 1$, then $T(f_1, \delta_1)$, or $T(f_{m+1}, \delta_{m+1})$, or $T(f_{m+2}, \delta_{m+2})$, or $T(f_{m+t+2}, \delta_{m+t+2})$ is empty table and $M(T, \bar{\delta}) \leq 1$. Let $\delta_1 = 1$, $\delta_{m+1} = 0$, $\delta_{m+2} = 1$ and $\delta_{m+t+2} = 0$. One can show that in this case there exist $i \in \{1, \ldots, m\}$ and $j \in \{m + 2, \ldots, m + t + 1\}$ such that $\delta_i = 1$, $\delta_{i+1} = 0$, $\delta_j = 1$, and $\delta_{j+1} = 0$. It is clear that the table $T(f_i, \delta_i)(f_{i+1}, \delta_{i+1})(f_j, \delta_j)(f_{j+1}, \delta_{j+1})$ contains exactly one row. So $M(T, \bar{\delta}) \leq 4$ and $M(T) \leq 4$. □

Using Theorem 3.11 we obtain

**Corollary 4.29**

$$L(T(S,\mu)) \leq 4 .$$

What can we do if we would like to solve the problem of minimization of test cardinality exactly? One of possible ways is to use so-called *Boolean reasoning* [80, 89].

Let $T$ be a decision table with $n$ columns labeled with attributes $f_1, \ldots, f_n$. Consider boolean variables $x_1, \ldots, x_n$. We will try to describe the notion of test for $T$ using these variables. The value of the variable $x_i$ is equal to 1 if and only if we include the attribute $f_i$ into the considered set. For any pair of rows with different decisions we can represent the fact that on the considered set of columns these two rows are different in the following way: $x_{i_1} \vee \ldots \vee x_{i_m} = 1$ where $f_{i_1}, \ldots, f_{i_m}$ are all columns on which the considered two rows are different. The formula of the kind $x_{i_1} \vee \ldots \vee x_{i_m}$ is called an *elementary disjunction*. Consider the conjunction of all such elementary disjunctions corresponding to pairs of rows with different decisions.

Let us multiply the elementary disjunctions (here the conjunction is considered as an analog of multiplication). After that we will use operations of absorption $A \vee A \wedge B = A$ and $A \wedge A = A \vee A = A$ while it is possible. As a result, we obtain a disjunction of *elementary conjunctions* of the kind $x_{j_1} \wedge \ldots \wedge x_{j_p}$. Each such conjunction $x_{j_1} \wedge \ldots \wedge x_{j_p}$ corresponds to a reduct $\{f_{j_1}, \ldots, f_{j_p}\}$, and the set of all conjunctions corresponds to the set of all reducts. Among the constructed reducts we can choose a reduct with minimum cardinality.

*Example 4.30.* Consider the table $T$ depicted in Fig. 4.6. Let us construct for this table a conjunction of elementary disjunctions (describing the notion of test for $T$) and transform it to a disjunction of elementary conjunctions corresponding to reducts of the table $T$. We will omit the symbol $\wedge$ in formulas: $(x_1 \vee x_3)(x_2 \vee x_3) = x_1x_2 \vee x_1x_3 \vee x_3x_2 \vee x_3x_3 = x_1x_2 \vee x_1x_3 \vee x_3x_2 \vee x_3 = x_1x_2 \vee x_3$. So, the considered table $T$ has exactly two reducts: $\{f_3\}$ and $\{f_1, f_2\}$. The set $\{f_3\}$ is a test with minimum cardinality for $T$.

$T =$

| $f_1$ | $f_2$ | $f_3$ | |
|---|---|---|---|
| 1 | 1 | 1 | 1 |
| 0 | 1 | 0 | 2 |
| 1 | 0 | 0 | 2 |

**Fig. 4.6**

Note that similar approach to construction of all reducts was considered in Sect. 2.2.2.

## 4.4 Conclusions

This chapter is devoted to the study of approximate and exact algorithms for the minimization of cardinality of tests, length of decision rules and depth of decision trees, and also for the optimization of decision rules systems.

The comparison of Theorems 4.3 and 4.11, 4.5 and 4.13, 4.7 and 4.15, 4.19 (Corollary 4.20) and 4.22 shows that, under the assumption $NP \not\subseteq DTIME(n^{O(\log \log n)})$, the considered in the chapter greedy algorithms are close (from the point of view of accuracy) to the best polynomial approximate algorithms for the minimization of test cardinality, rule length and tree depth, and also for the optimization of decision rule systems.

We found an interesting difference between rules and trees, and tests. If $P \neq NP$, then there are no algorithms for test cardinality minimization which are polynomial depending on the number of separable subtables and the number of attributes in the considered decision table. For trees and rules such algorithms exist.

All results considered in this chapter are true also (after small natural changes) for $k$-valued decision tables filled by numbers from the set $\{0, 1, \ldots, k-1\}$, $k \geq 3$.

Note also that information obtained during the work of greedy algorithms for construction of decision rules end tests can be used for creation of lower bounds on the minimum length of decision rules and minimum cardinality of tests (see [59]).

# 5

# Decision Tables with Many-Valued Decisions

Decision tables with many-valued decisions arise often in various applications. In contrast to decision tables with one-valued decisions, in decision tables with many-valued decisions each row is labeled with a nonempty finite set of natural numbers (decisions). If we want to find all decisions corresponding to a row, we deal with the same mathematical object as decision table with one-valued decisions: it is enough to code different sets of decisions by different numbers. However, if we want to find one (arbitrary) decision from the set attached to a row, we have essentially different situation.

In particular, in rough set theory [70, 80] decision tables are considered often that have equal rows labeled with different decisions. The set of decisions attached to equal rows is called the *generalized decision* for each of these equal rows. The usual way is to find for a given row its generalized decision. However, the problems of finding an arbitrary decision or one of the most frequent decisions from the generalized decision look also reasonable.

This chapter consists of ten sections. Section 5.1 contains examples of decision tables with many-valued decisions. In Sect. 5.2, main notions are discussed. In Sect. 5.3, relationships among decision trees, rules and tests are considered. In Sects. 5.4 and 5.5, lower and upper bounds on complexity of trees, rules and tests are studied. Approximate algorithms for optimization of tests, rules and trees are considered in Sects. 5.6 and 5.7. Section 5.8 is devoted to the discussion of exact algorithms for optimization of trees, rules and tests. Section 5.9 contains an example which illustrates the constructions considered in this chapter. Section 5.10 contains conclusions.

## 5.1 Examples Connected with Applications

Consider some examples of decision tables with many-valued decisions. In these examples, instead of numbers sometimes we will use letters in the capacity of decisions.

M. Moshkov and B. Zielosko: Combinatorial Machine Learning, SCI 360, pp. 69–86.
springerlink.com © Springer-Verlag Berlin Heidelberg 2011

*Example 5.1.* Let we have three inverted cups and a small ball under one of these cups (see Fig. 1.5). We should find a number of cup without ball. To this end, we will use the same attributes $f_1, f_2, f_3$ as in the example Three Cups and Small Ball (see Sect. 1.3.1). The decision table, corresponding to

| $f_1$ | $f_2$ | $f_3$ | |
|---|---|---|---|
| 1 | 0 | 0 | $\{2,3\}$ |
| 0 | 1 | 0 | $\{1,3\}$ |
| 0 | 0 | 1 | $\{1,2\}$ |



**Fig. 5.1** **Fig. 5.2**

this problem is represented in Fig. 5.1, and a decision tree for this problem solving is represented in Fig. 5.2. The decision rule system

$$\{f_1 = 0 \rightarrow 1, f_2 = 0 \rightarrow 2\}$$

is a complete decision rule system for the considered table.

*Example 5.2.* Let we have two real numbers $x$ and $y$. We should find a maximum number among $x$ and $y$. To this end, we will use two binary attributes $f_1$ and $f_2$: $f_1 = 0$ if and only if $x < y$, and $f_2 = 0$ if and only if $y < x$. Corresponding decision table is depicted in Fig. 5.3, and a decision tree for this table is depicted in Fig. 5.4. The decision rule system

$$\{f_1 = 1 \rightarrow x, f_2 = 1 \rightarrow y\}$$

is a complete decision rule system for the considered table.

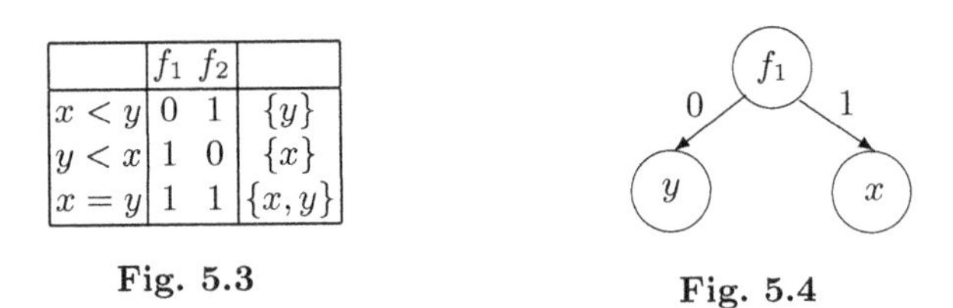

| | $f_1$ | $f_2$ | |
|---|---|---|---|
| $x < y$ | 0 | 1 | $\{y\}$ |
| $y < x$ | 1 | 0 | $\{x\}$ |
| $x = y$ | 1 | 1 | $\{x, y\}$ |



**Fig. 5.3** **Fig. 5.4**

*Example 5.3.* Consider now the combinatorial circuit $S$ depicted in Fig. 5.5. Each input of $S$ can work correctly or can have constant fault 1 (see the example Diagnosis of One-Gate Circuit in Sect. 1.3.2). Let we know that at least one such fault exists in $S$. We should find an input with fault. To this end, we can use attributes from the set $\{0, 1\}^3$. We give a tuple from this set on inputs of $S$ and observe the value of the output of $S$ which is the value of the considered attribute. It is clear that the circuit $S$ with at least one fault

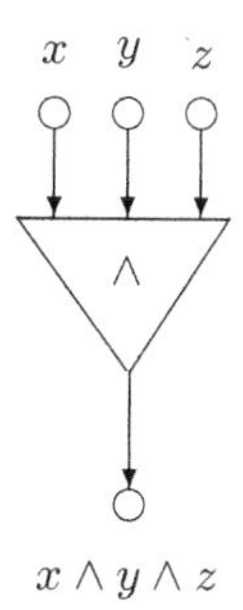


**Fig. 5.5**

1 on an input realizes one of functions from the set $\{1, x, y, z, xy, xz, yz\}$ (we write $xy$ instead of $x \wedge y$). Corresponding decision table is represented in Fig. 5.6.

| | 000 | 001 | 010 | 011 | 100 | 101 | 110 | 111 | |
|---|---|---|---|---|---|---|---|---|---|
| 1 | 1 | 1 | 1 | 1 | 1 | 1 | 1 | 1 | $\{x, y, z\}$ |
| $x$ | 0 | 0 | 0 | 0 | 1 | 1 | 1 | 1 | $\{y, z\}$ |
| $y$ | 0 | 0 | 1 | 1 | 0 | 0 | 1 | 1 | $\{x, z\}$ |
| $z$ | 0 | 1 | 0 | 1 | 0 | 1 | 0 | 1 | $\{x, y\}$ |
| $xy$ | 0 | 0 | 0 | 0 | 0 | 0 | 1 | 1 | $\{z\}$ |
| $xz$ | 0 | 0 | 0 | 0 | 0 | 1 | 0 | 1 | $\{y\}$ |
| $yz$ | 0 | 0 | 0 | 1 | 0 | 0 | 0 | 1 | $\{x\}$ |

**Fig. 5.6**

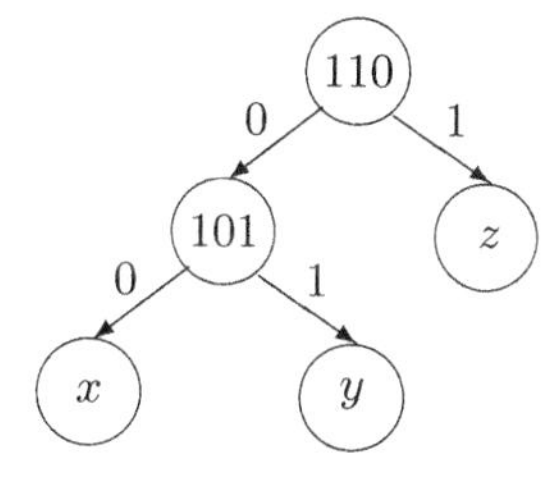


**Fig. 5.7**

In Fig. 5.7, we see a decision tree for this table (which solves the problem of searching for an input with fault). Each decision tree for this table must have at least 3 terminal nodes since there are three rows with pairwise disjoint sets of decisions $\{x\}$, $\{y\}$ and $\{z\}$. Thus, the decision tree depicted in Fig. 5.7 has minimum depth. A decision rule system

$$\{011 = 1 \rightarrow x, 101 = 1 \rightarrow y, 110 = 1 \rightarrow z\}$$

is a complete decision rule system for the considered table.

Decision tables with many-valued decisions arise often in different exactly formulated problems: in discrete optimization (see example Traveling Salesman Problem with Four Cities in Sect. 1.3.5 and Example 5.2), in fault diagnosis (Example 5.3), in computational geometry (example Problem of Three Post-Offices in Sect. 1.3.3). It is possible to point out on such problems in the area of pattern recognition.

However, the main source of decision tables with many-valued decisions is data tables with experimental data. It is possible that in such a table there are equal rows with different decisions. It is clear that for equal rows a constructed decision tree will give us the same decision. If we want to have minimum number of incorrect answers, we must choose a decision for which the number of the considered equal rows labeled with this decision is maximum. It is possible that there exist more than one such decision. In this case we obtain a decision table with many-valued decisions.

*Example 5.4.* Let we have the data table $D$ depicted in Fig. 5.8. All variables $x_1$, $x_2$, $x_3$ and $y$ are discrete. We must predict the value of $y$ using variables $x_1$, $x_2$, $x_3$ or, more exactly, values of attributes $f_1 = x_1$, $f_2 = x_2$, and $f_3 = x_3$. Corresponding decision table with many-valued decisions is depicted in Fig. 5.9.

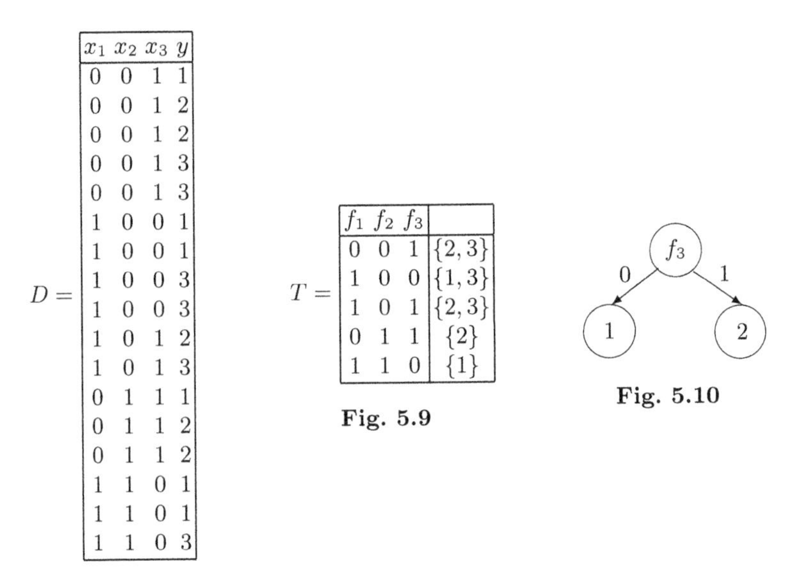

$D =$

| $x_1$ | $x_2$ | $x_3$ | $y$ |
|---|---|---|---|
| 0 | 0 | 1 | 1 |
| 0 | 0 | 1 | 2 |
| 0 | 0 | 1 | 2 |
| 0 | 0 | 1 | 3 |
| 0 | 0 | 1 | 3 |
| 1 | 0 | 0 | 1 |
| 1 | 0 | 0 | 1 |
| 1 | 0 | 0 | 3 |
| 1 | 0 | 0 | 3 |
| 1 | 0 | 1 | 2 |
| 1 | 0 | 1 | 3 |
| 0 | 1 | 1 | 1 |
| 0 | 1 | 1 | 2 |
| 0 | 1 | 1 | 2 |
| 1 | 1 | 0 | 1 |
| 1 | 1 | 0 | 1 |
| 1 | 1 | 0 | 3 |

**Fig. 5.8**

$T =$

| $f_1$ | $f_2$ | $f_3$ | |
|---|---|---|---|
| 0 | 0 | 1 | $\{2,3\}$ |
| 1 | 0 | 0 | $\{1,3\}$ |
| 1 | 0 | 1 | $\{2,3\}$ |
| 0 | 1 | 1 | $\{2\}$ |
| 1 | 1 | 0 | $\{1\}$ |

**Fig. 5.9**



**Fig. 5.10**

A decision tree for this decision table is represented in Fig. 5.10. A decision rule system

$$\{f_3 = 0 \rightarrow 1, f_3 = 1 \rightarrow 2\}$$

is a complete decision rule system for the considered table.

Note that there are other ways to form a set of decisions attached to a row: we can include to this set all decisions attached to equal rows, or the first $k$ most frequent decisions for equal rows, etc.

## 5.2 Main Notions

Now we consider formal definitions of notions corresponding to many-valued decision tables.

A *decision table with many-valued* decisions is a rectangular table $T$ filled by numbers from the set $\{0, 1\}$. Columns of this table are labeled with names of attributes $f_1, \ldots, f_n$. Rows of the table are pairwise different, and each row is labeled with a nonempty finite set of natural numbers (set of decisions). Note that each decision table with one-valued decisions can be interpreted also as a decision table with many-valued decisions. In such table each row is labeled with a set of decisions which has one element.

We will associate a *game* of two players with this table. The first player chooses a row of the table, and the second player must find a number from the set corresponding to this row. To this end, he can choose columns (attributes) and ask the first player what is at the intersection of the considered row and these columns.

The notion of a *decision tree over* $T$ coincides with the notion of a decision tree over a decision table with one-valued decisions.

We will say that a decision tree $\Gamma$ over the decision table $T$ is a *decision tree for* $T$ if for any row of $T$ the work of $\Gamma$ finishes in a terminal node which is labeled with a number from the set attached to the considered row.

A decision tree for the table $T$ can be interpreted as a right strategy of the second player in the considered game.

We denote by $h(T)$ the minimum depth of a decision tree for the table $T$.

The notions of a *decision rule over* $T$ and a decision rule *realizable* for a row $r$ of $T$ coincide with corresponding notions for decision tables with one-valued decisions.

A rule over $T$ with the right-hand side $t$ is called *true* for $T$ if for any row $r$ of $T$, such that this rule is realizable for the row $r$, the number $t$ belongs to the set of decisions attached to the row $r$. We denote by $L(T, r)$ the minimum length of a rule over $T$ which is true for $T$ and realizable for $r$.

A nonempty set $S$ of decision rules over $T$ is called a *complete decision rule system for* $T$ if each rule from $S$ is true for $T$, and for every row of $T$ there exists a rule from $S$ which is realizable for this row. We denote by $L(S)$ the maximum length of a rule from $S$, and by $L(T)$ we denote the minimum value of $L(S)$ among all complete decision rule systems $S$ for $T$.

We will say that $T$ is a *degenerate* table if either $T$ has no rows, or the intersection of sets of decisions attached to rows of $T$ is nonempty (in this case, we will say that rows of $T$ have *common* decision).

A *test for the table* $T$ is a subset of columns $\{f_{i_1}, \ldots, f_{i_m}\}$ such that for any numbers $\delta_1, \ldots, \delta_m \in \{0, 1\}$ the subtable $T(f_{i_1}, \delta_m) \ldots T(f_{i_m}, \delta_m)$ is a degenerate table. Empty set is a test for $T$ iff $T$ is a degenerate table. Note that for decision tables with one-valued decisions this notion is equivalent to the notion of the test considered earlier.

*A reduct* for the table $T$ is a test for $T$ for which each proper subset is not a test. It is clear that each test has a reduct as a subset. We denote by $R(T)$ the minimum cardinality of a reduct for $T$.

## 5.3 Relationships among Decision Trees, Rules and Tests

**Theorem 5.5.** *Let $T$ be a decision table with many-valued decisions.*

*a) If $\Gamma$ is a decision tree for $T$ then the set of attributes attached to working nodes of $\Gamma$ is a test for the table $T$.*
*b) Let $\{f_{i_1}, \ldots, f_{i_m}\}$ be a test for $T$. Then there exists a decision tree for $T$ which uses only attributes from $\{f_{i_1}, \ldots, f_{i_m}\}$ and for which $h(\Gamma) = m$.*

*Proof.* Let $T$ have $n$ columns labeled with attributes $f_1, \ldots, f_n$.

a) Let $\Gamma$ be a decision tree for the table $T$. Let, for simplicity, $\{f_1, \ldots, f_t\}$ be the set of attributes attached to working nodes of $\Gamma$. Let $(\delta_1, \ldots, \delta_t) \in \{0,1\}^t$. We show that $T' = T(f_1, \delta_1) \ldots (f_t, \delta_t)$ is a degenerate table. Consider a path in $\Gamma$ from the root to a terminal node $v$ which satisfies the following condition: let $f_{i_1}, \ldots, f_{i_m}$ be attributes attached to working nodes of this path. Then the edges of this path are labeled with numbers $\delta_{i_1}, \ldots, \delta_{i_m}$. Consider the table $T'' = T(f_{i_1}, \delta_{i_1}) \ldots (f_{i_m}, \delta_{i_m})$. It is clear that the set of rows of $T''$ coincides with the set of rows of $T$ for which the work of $\Gamma$ finishes in the node $v$. Since $\Gamma$ is a decision tree for $T$, the table $T''$ is degenerate. It is clear that $T'$ is a subtable of $T''$. Therefore $T'$ is a degenerate subtable of $T$. Taking into account that $(\delta_1, \ldots, \delta_t)$ is an arbitrary tuple from $\{0,1\}^t$ we obtain $\{f_1, \ldots, f_t\}$ is a test for the table $T$.
b) Let $\{f_{i_1}, \ldots, f_{i_m}\}$ be a test for the table $T$. Consider a decision tree $\Gamma$ which consists of $(m+1)$ layers. For $j = 1, \ldots, m$, all nodes of the $j$-th layer are working nodes labeled with the attribute $f_{i_j}$. All nodes from $(m+1)$-th layer are terminal nodes. Let $v$ be an arbitrary terminal node, and let the edges in the path from the root to $v$ be labeled with numbers $\delta_1, \ldots, \delta_m$. Denote $T(v) = T(f_{i_1}, \delta_1) \ldots (f_{i_m}, \delta_m)$. Since $\{f_{i_1}, \ldots, f_{i_m}\}$ is a test for $T$, the table $T(v)$ is degenerate. If there are no rows in $T(v)$ then the node $v$ is labeled with the number 1. If $T(v)$ has nodes, and $d$ is a common decision for all rows of $T(v)$ then $v$ is labeled with $d$. It is clear that $\Gamma$ is a decision tree for $T$, and $h(\Gamma) = m$. □

**Corollary 5.6.** *Let $T$ be a decision table with many-valued decisions. Then*

$$h(T) \leq R(T) .$$

**Theorem 5.7.** *Let $T$ be a decision table with many-valued decisions containing $n$ columns labeled with attributes $f_1, \ldots, f_n$.*

*1. If $S$ is a complete system of decision rules for $T$ then the set of attributes from rules in $S$ is a test for $T$.*
*2. If $F = \{f_{i_1}, \ldots, f_{i_m}\}$ is a test for $T$ then there exists a complete system $S$ of decision rules for $T$ which uses only attributes from $F$ and for which $L(S) = m$.*

*Proof.* 1. Let $S$ be a complete system of decision rules for $T$. Let, for simplicity, $\{f_1, \ldots, f_t\}$ be the set of attributes from rules in $S$, and $(\delta_1, \ldots, \delta_t) \in \{0,1\}^t$. We show that $T' = T(f_1, \delta_1) \ldots (f_t, \delta_t)$ is a degenerate table. If $T'$ has no rows, then $T'$ is a degenerate table. Let $T'$ have at least one row $\bar{\delta} = (\delta_1, \ldots, \delta_t, \delta_{t+1}, \ldots, \delta_n)$. Since $S$ is a complete system of decision rules for $T$, there is a rule

$$f_{i_1} = \delta_{i_1} \wedge \ldots \wedge f_{i_m} = \delta_{i_m} \rightarrow d$$

in $S$ which is realizable for $\bar{\delta}$ and true for $T$. Consider the table $T'' = T(f_{i_1}, \delta_{i_1}) \ldots (f_{i_m}, \delta_{i_m})$. It is clear that the set of rows of $T''$ coincides with the set of rows of $T$ for which the considered rule is realizable. Since this rule is true for $T$, the set of rows of $T''$ has common decision $d$. It is clear that $T'$ is a subtable of $T''$. Therefore $T'$ is a degenerate table. Taking into account that $(\delta_1, \ldots, \delta_t)$ is an arbitrary tuple from $\{0,1\}^t$ we obtain $\{f_1, \ldots, f_t\}$ is a test for the table $T$.

2. Let $F = \{f_{i_1}, \ldots, f_{i_m}\}$ be a test for the table $T$. For each $\bar{\delta} = (\delta_1, \ldots, \delta_m) \in \{0,1\}^m$ such that the subtable $T(\bar{\delta}) = T(f_{i_1}, \delta_1) \ldots (f_{i_m}, \delta_m)$ is nonempty, we construct a decision rule

$$f_{i_1} = \delta_1 \wedge \ldots \wedge f_{i_m} = \delta_m \rightarrow d ,$$

where $d$ is a common decision for the set of rows of $T(\bar{\delta})$. Such common decision exists since $F$ is a test for $T$. We denote by $S$ the set of constructed rules. It is clear that each rule from $S$ is true for $T$, and for each row of $T$ there exists a rule in $S$ which is realizable for this row. Thus, $S$ is a complete decision rule system for $T$, $L(S) = m$ and $S$ uses only attributes from $F$. □

**Corollary 5.8.** $L(T) \leq R(T)$.

**Theorem 5.9.** *Let $\Gamma$ be a decision tree for $T$, and $S$ be the set of decision rules corresponding to paths in $\Gamma$ from the root to terminal nodes. Then $S$ is a complete system of decision rules for $T$ and $L(S) = h(\Gamma)$.*

*Proof.* Since $\Gamma$ is a decison tree for $T$, for each row $r$ of $T$ there exists a path $\tau$ from the root to a terminal node $v$ of $\Gamma$ such that the work of $\Gamma$ for $r$ finishes in $v$, and $v$ is labeled with a decision $t$ which belongs to the set of decisions attached to $r$. It is clear that the rule rule$(\tau)$ corresponding to the path $\tau$ is realizable for $r$. It is clear also that for each row $r'$ of $T$ such that the rule rule$(\tau)$ is realizable for $r'$, the set of decisions attached to $r'$ contains the decision $t$. So rule$(\tau)$ is true for $T$. It is clear that the length of rule$(\tau)$ is equal to the length of path $\tau$. Therefore $S$ is a complete decision rule system for $T$ and $L(S) = h(\Gamma)$. □

**Corollary 5.10.** $L(T) \leq h(T)$.

## 5.4 Lower Bounds

From Corollaries 5.6 and 5.10 it follows that $L(T) \leq h(T) \leq R(T)$. So each lower bound on $L(T)$ is also a lower bound on $h(T)$ and $R(T)$, and each lower bound on $h(T)$ is also a lower bound on $R(T)$. Example which illustrates the obtained bounds can be found in Sect. 5.9.

Let $T$ be a nonempty decision table with many-valued decisions. A nonempty finite set $B$ of natural numbers is called a *system of representatives for the table* $T$ if for each row of $T$ the set of decisions attached to this row has a nonempty intersection with $B$. We denote by $S(T)$ the minimum cardinality of a system of representatives for the table $T$.

**Theorem 5.11.** *Let* $T$ *be a nonempty decision table with many-valued decisions. Then*

$$h(T) \geq \log_2 S(T) .$$

The proof of this theorem coincides with the proof of Theorem 3.1 if instead of the parameter $D(T)$ we will use the parameter $S(T)$.

**Theorem 5.12.** *Let* $T$ *be a decision table. Then*

$$h(T) \geq \log_2(R(T) + 1) .$$

Proof of this statement coincides with the proof of Theorem 3.2, but instead of Theorem 2.23 we must use Theorem 5.5.

Let $T$ be a decision table with many-valued decisions, which has $n$ columns labeled with attributes $\{f_1, \ldots, f_n\}$. We will use the following definition of the parameter $M(T)$. If $T$ is a degenerate table then $M(T) = 0$. Let now $T$ be a nondegenerate table. Let $\bar{\delta} = (\delta_1, \ldots, \delta_n) \in \{0, 1\}^n$. Then $M(T, \bar{\delta})$ is the minimum natural $m$ such that there exist attributes $f_{i_1}, \ldots, f_{i_m} \in \{f_1, \ldots, f_n\}$ for which $T(f_{i_1}, \delta_{i_1}) \ldots (f_{i_m}, \delta_{i_m})$ is a degenerate table. We denote $M(T) = \max\{M(T, \bar{\delta}) : \bar{\delta} \in \{0, 1\}^n\}$.

**Lemma 5.13.** *Let* $T$ *be a decision table with many-valued decisions, and* $T'$ *be a subtable of* $T$*. Then*

$$M(T) \geq M(T') .$$

Proof of this lemma coincides with the proof of Lemma 3.4.

**Theorem 5.14.** *Let* $T$ *be a decision table with many-valued decisions. Then*

$$h(T) \geq M(T) .$$

The proof of this theorem coincides with the proof of Theorem 3.6.

Let $T$ be a decision table with many-valued decisions having $n$ columns, and let $m \leq n$. The notions of a *$(T, m)$-proof-tree* and a *proof-tree for the bound $h(T) \geq m$ for the table $T$* are the same as for decisions tables with one-valued decisions.

**Theorem 5.15.** *Let $T$ be a nondegenerate decision table with many-valued decisions having $n$ columns, and $m$ be a natural number such that $m \leq n$. Then a proof-tree for the bound $h(T) \geq m$ exists if and only if the inequality $h(T) \geq m$ holds.*

The proof of this theorem coincides with the proof of Theorem 3.9.

**Theorem 5.16.** *Let $T$ be a decision table with many-valued decisions and $\Delta(T)$ be the set of rows of $T$. Then $L(T, \bar{\delta}) = M(T, \bar{\delta})$ for any row $\bar{\delta} \in \Delta(T)$ and $L(T) = \max\{M(T, \bar{\delta}) : \bar{\delta} \in \Delta(T)\}$.*

*Proof.* Let $T$ have $n$ columns labeled with attributes $f_1, \ldots, f_n$, and $\bar{\delta} = (\delta_1, \ldots, \delta_n)$ be a row of $T$. One can show that a decision rule

$$f_{i_1} = b_1 \wedge \ldots \wedge f_{i_m} = b_m \rightarrow d$$

is true for $T$ and realizable for $\bar{\delta}$ if and only if $b_1 = \delta_{i_1}, \ldots, b_m = \delta_{i_m}$ and $d$ is a common decision for the set of rows of the table $T(f_{i_1}, b_1) \ldots (f_{i_m}, b_m)$. From here it follows that $L(T, \bar{\delta}) = M(T, \bar{\delta})$ and $L(T) = \max\{M(T, \bar{\delta}) : \bar{\delta} \in \Delta(T)\}$.

## 5.5 Upper Bounds

We know that $L(T) \leq h(T) \leq R(T)$. Therefore each upper bound on $R(T)$ is also an upper bound on $h(T)$ and $L(T)$, and each upper bound on $h(T)$ is also an upper bound on $L(T)$. Example which illustrates these bounds can be found in Sect. 5.9.

**Theorem 5.17.** *Let $T$ be a decision table with many-valued decisions. Then*

$$R(T) \leq N(T) - 1 .$$

*Proof.* We prove this inequality by induction on $N(T)$. If $N(T) = 1$ then, evidently, $R(T) = 0$. Let $m \geq 1$ and for any decision table with many-valued decisions having at most $m$ rows the inequality $R(T) \leq N(T) - 1$ holds.

Let $T$ be a decision table with many-valued decisions for which $N(T) = m + 1$. Let us prove that $R(T) \leq m$. Since $T$ has at least two rows, and rows of $T$ are pairwise different, there exists a column $f_i$ of $T$, which has both 0 and 1. Let us consider subtables $T(f_i, 0)$ and $T(f_i, 1)$. It is clear that each of these subtables has at most $m$ rows. Using inductive hypothesis we obtain that for any $\delta \in \{0, 1\}$ there exists a test $B_\delta$ for the table $T(f_i, \delta)$ such that $|B_\delta| \leq N(T(f_i, \delta)) - 1$. We denote $B = \{f_i\} \cup B_0 \cup B_1$. Let us show that $B$ is a test for the table $T$.

Let, for the definiteness, $B = \{f_1, \ldots, f_m\}$, $B_0 = \{f_{k_1}, \ldots, f_{k_t}\}$ and $B_1 = \{f_{j_1}, \ldots, f_{j_p}\}$. Consider an arbitrary tuple $(\delta_1, \ldots, \delta_m) \in \{0, 1\}^m$. If $\delta_i = 0$ then, evidently, $T(f_i, 0)(f_{k_1}, \delta_{k_1}) \ldots (f_{k_t}, \delta_{k_t})$ is degenerate. If $\delta_i = 1$ then, evidently, $T(f_i, 1)(f_{j_1}, \delta_{j_1}) \ldots (f_{j_p}, \delta_{j_p})$ is degenerate. Therefore

$$T(f_1, \delta_1) \ldots (f_m, \delta_m)$$

is degenerate. Hence $B$ is a test for $T$.

Since $N(T) = N(T(f_i, 0)) + N(T(f_i, 1))$, we have $|B| \leq 1 + N(T(f_i, 0)) - 1 + N(T(f_i, 1)) - 1 = N(T) - 1$. □

**Theorem 5.18.** *Let $T$ be a decision table with many-valued decisions. Then*

$$h(T) \leq M(T) \log_2 N(T) .$$

Proof of this theorem coincides with the proof of Theorem 3.17.

A nonempty decision table $T$ with many-valued decisions will be called a *diagnostic* table if rows of this table are labeled with pairwise disjoint sets of decisions.

**Corollary 5.19.** *Let $T$ be a diagnostic table with many-valued decisions. Then*

$$\max\{M(T), \log_2 N(T)\} \leq h(T) \leq M(T) \log_2 N(T) .$$

Note, that for diagnostic table $T$ with many-valued decisions $S(T) = N(T)$.

## 5.6 Approximate Algorithms for Optimization of Tests and Decision Rules

In this section, we consider approximate polynomial algorithms for problem of minimization of test cardinality, problem of minimization of decision rule length, and problem of optimization of decision rule system. Corresponding examples can be found in Sect. 5.9.

### 5.6.1 Optimization of Tests

Let $T$ be a decision table with many-valued decisions. A subtable $T'$ of $T$ is called *boundary* subtable if $T'$ is not degenerate but each proper subtable of $T'$ is degenerate.

We denote by $B(T)$ the number of boundary subtables of the table $T$. We denote by $Tab(t)$, where $t$ is a natural number, the set of decision tables with many-valued decisions such that each row in the table has at most $t$ decisions (labeled with a set of decisions which cardinality is at most $t$).

**Proposition 5.20.** *Let $T'$ be a boundary subtable with $m$ rows. Then each row of $T'$ is labeled with a set of decisions which cardinality is at least $m - 1$.*

*Proof.* Let rows of $T'$ be labeled with sets of decisions $D_1, \ldots, D_m$ respectively. Then $D_1 \cap \ldots \cap D_m = \emptyset$ and for any $i \in \{1, \ldots, m\}$, the set $D_1 \cap \ldots \cap D_{i-1} \cap D_{i+1} \cap \ldots \cap D_m$ contains a number $d_i$. Assume that $i \neq j$ and $d_i = d_j$. Then $D_1 \cap \ldots \cap D_m \neq \emptyset$ which is impossible. Therefore $d_1, \ldots, d_m$ are pairwise different numbers. It is clear that for $i = 1, \ldots, m$, the set $\{d_1, \ldots, d_m\} \setminus \{d_i\}$ is a subset of the set $D_i$. □

**Corollary 5.21.** *Each boundary subtable of a table $T \in Tab(t)$ has at most $t + 1$ rows.*

Therefore, for tables from $Tab(t)$, there exists a polynomial algorithm for computation of the parameter $B(T)$, and for construction of the set of boundary subtables of the table $T$. For example, for any decision table $T$ with one-valued decisions (for any table from $Tab(1)$) the equality $B(T) = P(T)$ holds.

Let $T$ be a decision table with many-valued decisions. It is clear that $T$ is a degenerate table if and only if $B(T) = 0$.

Let us consider an algorithm which for a given decision table with many-valued decisions $T$ constructs a test for $T$. Let $T$ contain $n$ columns labeled with attributes $f_1, \ldots, f_n$. We construct a set cover problem $A(T)$, $F(T)$ corresponding to the table $T$, where $A(T)$ is the set of all boundary subtables of $T$, $F(T) = \{S_1, \ldots, S_n\}$, and, for $i = 1, \ldots, n$, $S_i$ is the set of boundary subtables from $A(T)$ in each of which there exists a pair of rows that are different in the column $f_i$. One can show that $\{f_{i_1}, \ldots, f_{i_m}\}$ is a test for $T$ if and only if $\{S_{i_1}, \ldots, S_{i_m}\}$ is a cover for $A(T)$, $F(T)$. Let us apply the greedy algorithm for set cover problem to $A(T)$, $F(T)$. As a result, we obtain a cover corresponding to a test for $T$. This test is a result of the considered algorithm work. We denote by $R_{\text{greedy}}(T)$ the cardinality of the constructed test.

**Theorem 5.22.** *Let $T$ be a decision table with many-valued decisions. Then*

$$R_{\text{greedy}}(T) \leq R(T) \ln B(T) + 1 .$$

This result follows immediately from the description of the considered algorithm and from Theorem 4.1.

For any natural $t$, for tables from the class $Tab(t)$ the considered algorithm has polynomial time complexity.

**Proposition 5.23.** *The problem of minimization of test cardinality for decision tables with many-valued decisions is $NP$-hard.*

This result follows immediately from Proposition 4.10.

**Theorem 5.24.** *If $NP \not\subseteq DTIME(n^{O(\log \log n)})$ then for any $\varepsilon$, $0 < \varepsilon < 1$, there is no polynomial algorithm which for a given decision table $T$ with many-valued decisions constructs a test for $T$ which cardinality is at most*

$$(1 - \varepsilon) R(T) \ln B(T) .$$

The considered theorem follows directly from Theorem 4.11.

### 5.6.2 Optimization of Decision Rules

We can apply greedy algorithm for set cover problem to construct decision rules for decision tables with many-valued decisions.

Let $T$ be a nondegenerate table with many-valued decisions containing $n$ columns labeled with attributes $f_1, \ldots, f_n$. Let $r = (b_1, \ldots, b_n)$ be a row of $T$, $D(r)$ be the set of decisions attached to $r$ and $d \in D(r)$.

We consider a set cover problem $A(T, r, d)$, $F(T, r, d) = \{S_1, \ldots, S_n\}$, where $A(T, r, d)$ is the set of all rows $r'$ of $T$ such that $d \notin D(r')$. For $i = 1, \ldots, n$, the set $S_i$ coincides with the set of all rows from $A(T, r, d)$ which are different from $r$ in the column $f_i$. One can show that the decision rule

$$f_{i_1} = b_{i_1} \wedge \ldots \wedge f_{i_m} = b_{i_m} \rightarrow d$$

is true for $T$ (it is clear that this rule is realizable for $r$) if and only if the subfamily $\{S_{i_1}, \ldots, S_{i_m}\}$ is a cover for the set cover problem $A(T, r, d)$, $F(T, r, d)$.

We denote $P(T, r, d) = |A(T, r, d)|$ and $L(T, r, d)$ the minimum length of a decision rule over $T$ which is true for $T$, realizable for $r$ and has $d$ on the right-hand side. It is clear that for the constructed set cover problem $C_{\min} = L(T, r, d)$.

Let us apply the greedy algorithm to the set cover problem $A(T, r, d)$, $F(T, r, d)$. This algorithm constructs a cover which corresponds to a decision rule rule$(T, r, d)$ which is true for $T$, realizable for $r$ and has $d$ on the right-hand side. We denote by $L_{\text{greedy}}(T, r, d)$ the length of rule$(T, r, d)$. From Theorem 4.1 it follows that

$$L_{\text{greedy}}(T, r, d) \leq L(T, r, d) \ln P(T, r, d) + 1 \, .$$

We denote by $L_{\text{greedy}}(T, r)$ the length of the rule constructed by the following polynomial algorithm (we will say about this algorithm as about modified greedy algorithm). For a given decision table $T$ with many-valued decisions and row $r$ of $T$, for each $d \in D(r)$ we construct the set cover problem $A(T, r, d)$, $F(T, r, d)$ and then apply to this problem the greedy algorithm. We transform the constructed cover to the rule rule$(T, r, d)$. Among the rules rule$(T, r, d)$, $d \in D(r)$, we choose a rule with minimum length. This rule is the output of considered algorithm. We have

$$L_{\text{greedy}}(T, r) = \min\{L_{\text{greedy}}(T, r, d) : d \in D(r)\} \, .$$

It is clear that

$$L(T, r) = \min\{L(T, r, d) : d \in D(r)\} \, .$$

Let $K(T, r) = \max\{P(T, r, d) : d \in D(r)\}$.

Then

$$L_{\text{greedy}}(T, r) \leq L(T, r) \ln K(T, r) + 1 \, .$$

So we have the following statement.

**Theorem 5.25.** *Let $T$ be a nondegenerate decision table with many-valued decisions and $r$ be a row of $T$. Then*

$$L_{\text{greedy}}(T, r) \leq L(T, r) \ln K(T, r) + 1 \, .$$

We can use the considered modified greedy algorithm to construct a complete decision rule system for the decision table $T$ with many-valued decisions. To this end, we apply this algorithm sequentially to the table $T$ and to each row $r$ of $T$. As a result, we obtain a system of rules $S$ in which each rule is true for $T$ and for every row of $T$ there exists a rule from $S$ which is realizable for this row.

We denote $L_{\text{greedy}}(T) = L(S)$ and

$$K(T) = \max\{K(T,r) : r \in \Delta(T)\} ,$$

where $\Delta(T)$ is the set of rows of $T$. It is clear that $L(T) = \max\{L(T,r) : r \in \Delta(T)\}$. Using Theorem 5.25 we obtain

**Theorem 5.26.** *Let $T$ be a nondegenerate decision table with many-valued decisions. Then*

$$L_{\text{greedy}}(T) \leq L(T) \ln K(T) + 1 .$$

**Proposition 5.27.** *The problem of minimization of decision rule length for decision tables with many-valued decisions is $NP$-hard.*

This result follows immediately from Proposition 4.12.

**Theorem 5.28.** *If $NP \not\subseteq DTIME(n^{O(\log \log n)})$ then for any $\varepsilon$, $0 < \varepsilon < 1$, there is no polynomial algorithm that for a given nondegenerate decision table $T$ with many-valued decisions and row $r$ of $T$ constructs a decision rule which is true for $T$, realizable for $r$, and which length is at most*

$$(1 - \varepsilon)L(T,r) \ln K(T,r) .$$

The considered theorem follows directly from Theorem 4.13.

**Proposition 5.29.** *The problem of optimization of decision rule system is $NP$-hard.*

This result follows immediately from Proposition 4.14.

**Theorem 5.30.** *If $NP \not\subseteq DTIME(n^{O(\log \log n)})$ then for any $\varepsilon$, $0 < \varepsilon < 1$, there is no polynomial algorithm that for a given nondegenerate decision table $T$ with many-valued decisions constructs a decision rule system $S$ for $T$ such that*

$$L(S) \leq (1 - \varepsilon)L(T) \ln K(T) .$$

This theorem follows immediately from Theorem 4.15.

## 5.7 Approximate Algorithms for Decision Tree Optimization

Now we consider modified algorithm $U$ which for a given decision table with many-valued decisions $T$ constructs a decision tree $U(T)$ for the table $T$. The unique modification is the following: instead of $P(T)$ we will consider the parameter $B(T)$.

**Theorem 5.31.** *Let $T$ be a decision table with many-valued decisions. Then during the construction of the tree $U(T)$ the algorithm $U$ makes at most $2N(T) + 1$ steps.*

The proof of this theorem coincides with the proof of Theorem 4.17. From this theorem it follows that for any natural $t$ the algorithm $U$ has polynomial time complexity on the set $Tab(t)$. An example of this algorithm use can be found in Sect. 5.9.

**Lemma 5.32.** *Let $T$ be a decision table with many-valued decisions, $T'$ be a subtable of the table $T$, $f_i$ be an attribute attached to a column of $T$, and $\delta \in \{0, 1\}$. Then*

$$B(T) - B(T(f_i, \delta)) \geq B(T') - B(T'(f_i, \delta)) .$$

*Proof.* Denote by $J$ (respectively by $J'$) the set of boundary subtables of $T$ (respectively of $T'$) in each of which at least one row has at the intersection with column $f_i$ a number which is not equal to $\delta$. One can show that $J' \subseteq J$, $|J'| = B(T') - B(T'(f_i, \delta))$ and $|J| = B(T) - B(T(f_i, \delta))$. □

**Theorem 5.33.** *Let $T$ be a nondegenerate decision table with many-valued decisions. Then*

$$h(U(T)) \leq M(T) \ln B(T) + 1 .$$

If in the proof of Theorem 4.19 instead of the parameter $P$ we will use the parameter $B$, then we obtain a proof of Theorem 5.33.

Using Theorem 5.14 we obtain the following

**Corollary 5.34.** *For any nondegenerate decision table $T$ with many-valued decisions*

$$h(U(T)) \leq h(T) \ln B(T) + 1 .$$

**Proposition 5.35.** *The problem of minimization of decision tree depth for decision tables with many-valued decisions is $NP$-hard.*

This proposition follows immediately from Proposition 4.21, and the next theorem follows directly from Theorem 4.22.

**Theorem 5.36.** *If $NP \not\subseteq DTIME(n^{O(\log \log n)})$ then for any $\varepsilon > 0$ there is no polynomial algorithm which for a given nondegenerate decision table $T$ with many-valued decisions constructs a decision tree for $T$ which depth is at most*

$$(1 - \varepsilon) h(T) \ln B(T) .$$

## 5.8 Exact Algorithms for Optimization of Trees, Rules and Tests

The algorithm $W$ can be applied to an arbitrary decision table $T$ with many-valued decisions practically without any modification. Note only, that in the case when the subtable $D$ is degenerate, in the capacity of $d$ we should choose a common decision for all rows of $D$.

From Theorem 4.23 the following statement follows

**Theorem 5.37.** *For any nondegenerate decision table $T$ with many-valued decisions, the algorithm $W$ constructs a decision tree $W(T)$ for the table $T$ such that $h(W(T)) = h(T)$, and makes exactly $2|SEP(T)|+3$ steps. The time of the algorithm $W$ work is bounded from below by $|SEP(T)|$, and bounded from above by a polynomial on $|SEP(T)|$ and the number of columns in the table $T$.*

The same situation is with the algorithm $V$. We can almost repeat the proof of Lemma 4.25 and prove

**Lemma 5.38.** *For any table $D \in SEP(T)$ and any row $r$ of $D$ the decision rule attached to $r$ after the end of algorithm $V$ work is true for $D$, realizable for $r$, and has minimum length $L(D, r)$.*

Using this lemma it is not difficult to prove the following statement:

**Theorem 5.39.** *For any nondegenerate decision table $T$ with many-valued decisions and any row $r$ of $T$, the algorithm $V$ constructs a decision rule $V(T, r)$ which is true for $T$, realizable for $r$ and has minimum length $L(T, r)$. During the construction of optimal rules for rows of $T$ the algorithm $V$ makes exactly $2|SEP(T)| + 3$ steps. The time of the algorithm $V$ work is bounded from below by $|SEP(T)|$, and bounded from above by a polynomial on $|SEP(T)|$ and on the number of columns in the table $T$.*

From Theorem 4.27 the next statement follows immediately:

**Theorem 5.40.** *If $P \neq NP$ then there is no algorithm which for a given decision table $T$ with many-valued decisions constructs a test for $T$ with minimum cardinality, and for which the time of work is bounded from above by a polynomial depending on the number of columns in $T$ and the number of separable subtables of $T$.*

## 5.9 Example

Let us consider the decision table with many-valued decisions depicted in Fig. 5.11. One can show that this table has exactly two tests $\{f_1, f_3\}$ and $\{f_1, f_2, f_3\}$, and exactly one reduct $\{f_1, f_3\}$. Therefore $R(T) = 2$. It is not difficult to show that $\{1, 2\}$ is the minimum system of representatives for this table. Therefore $S(T) = 2$.

| | $f_1$ | $f_2$ | $f_3$ | |
|---|---|---|---|---|
| 1 | 1 | 1 | 1 | $\{1\}$ |
| 2 | 0 | 1 | 0 | $\{1,3\}$ |
| 3 | 1 | 1 | 0 | $\{2\}$ |
| 4 | 0 | 0 | 1 | $\{2,3\}$ |
| 5 | 1 | 0 | 0 | $\{1,2\}$ |

**Fig. 5.11**

Let us evaluate the value $M(T)$. One can show that $M(T,(1,1,1)) > 1$. Using the fact that $\{f_1, f_3\}$ is a test for $T$, we obtain $T(f_1,\delta_1)(f_3,\delta_3)$ is a degenerate table for any tuple $\bar{\delta} = (\delta_1,\delta_2,\delta_3) \in \{0,1\}^3$. Therefore $M(T,\bar{\delta}) \leq 2$ for any $\bar{\delta} \in \{0,1\}^3$. Thus, $M(T) = 2$ and $\max\{M(T,\bar{\delta}) : \bar{\delta} \in \Delta(T)\} = 2$.

So, we have the following lower bounds: $h(T) \geq \log_2 S(T) = 1$, $h(T) \geq \log_2(R(T)+1) = \log_2 3$ and $h(T) \geq M(T) = 2$, and an exact bound $L(T) = 2$. A proof-tree for the bound $h(T) \geq 2$ is represented in Fig. 5.12.

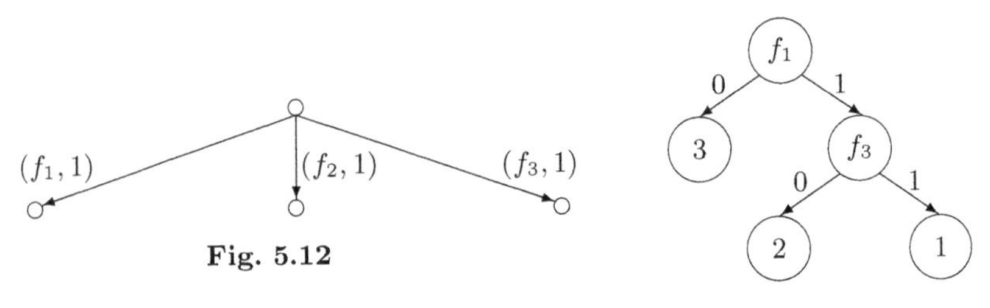


**Fig. 5.12**

**Fig. 5.13**

It is clear that $N(T) = 5$. So we have the following upper bounds on the value $h(T)$: $h(T) \leq R(T) = 2$, $h(T) \leq R(T) \leq N(T) - 1 = 4$ and $h(T) \leq M(T)\log_2 N(T) = 2\log_2 5$. Thus, $h(T) = 2$. A decision tree for $T$ which depth is equal to two is depicted in Fig. 5.13.

Let us find the set of boundary subtables of the table $T$. It is clear that $T \in Tab(2)$. From Corollary 5.21 it follows that each boundary subtable of the table $T$ contains at most three rows. From Proposition 5.20 it follows that if $T'$ is a boundary subtable of $T$ containing three rows then each row of $T'$ is labeled with a set of decision having at least two elements. We have exactly one such subtable $\{2,4,5\}$. Here 2, 4 and 5 are numbers of rows of the table $T$ which form the considered subtable. It is clear that $\{2,4,5\}$ is a boundary subtable of the table $T$. Each other boundary subtable of the table $T$ has exactly two rows. There are three such subtables: $\{1,3\}$, $\{1,4\}$, and $\{2,3\}$.

Let us apply the greedy algorithm for set cover problem for construction of a test for the table $T$. First, we transform the table $T$ into a set cover problem $A(T)$, $F(T) = \{S_1, S_2, S_3\}$, where $A(T)$ is the set of boundary subtables of $T$, and, for $i = 1,2,3$, the set $S_i$ coincides with the set of all subtables from

$A(T)$ in each of which at least two rows are different in the column $f_i$. We have $A(T) = \{\{2,4,5\},\{1,3\},\{1,4\},\{2,3\}\}$, $S_1 = \{\{2,4,5\},\{1,4\},\{2,3\}\}$, $S_2 = \{\{2,4,5\},\{1,4\}\}$, and $S_3 = \{\{2,4,5\},\{1,3\}\}$.

Now, we can apply the greedy algorithm for set cover problem to the problem $A(T)$, $F(T)$. As a result we obtain the cover $\{S_1, S_3\}$ and corresponding test $\{f_1, f_3\}$.

Let us apply the greedy algorithm for set cover problem for construction of a decision rule which is true for $T$ and realizable for $r$, where $r$ is the second row of $T$. For each $d \in D(r) = \{1,3\}$ we construct the set cover problem $A(T,r,d)$, $F(T,r,d) = \{S_1, S_2, S_3\}$, where $A(T,r,d)$ is the set of all rows $r'$ of $T$ such that $d \notin D(r')$, and $S_i$ coincides with the set of rows from $A(T,r,d)$ which are different from $r$ in the column $f_i$, $i = 1,2,3$. We have $A(T,r,1) = \{3,4\}$ (3 and 4 are numbers of rows), $F(T,r,1) = \{S_1 = \{3\}$, $S_2 = \{4\}$, $S_3 = \{4\}\}$, and $A(T,r,3) = \{1,3,5\}$, $F(T,r,3) = \{S_1 = \{1,3,5\}$, $S_2 = \{5\}$, $S_3 = \{1\}\}$. Now, we apply the greedy algorithm for set cover problem to each of the constructed set cover problems, and transform the obtained covers into decision rules. For the case $d = 1$, we obtain the cover $\{S_1, S_2\}$ and corresponding decision rule $f_1 = 0 \wedge f_2 = 1 \rightarrow 1$. For the case $d = 3$, we obtain the cover $\{S_1\}$ and corresponding decision rule $f_1 = 0 \rightarrow 3$. We choose the shortest rule $f_1 = 0 \rightarrow 3$ which is the result of our algorithm work.

Let us apply the greedy algorithm $U$ to the construction of a decision tree $U(T)$ for the table $T$. After the first step, we will have the tree which consists of one node labeled with the table $T$. Let us describe the second step. The table $T$ is not degenerate. So, for $i = 1,2,3$, we compute the value

$$Q(f_i) = \max\{B(T(f_i,0)), B(T(f_i,1))\} .$$

One can show that $Q(f_1) = \max\{0,1\} = 1$, $Q(f_2) = \max\{0,2\} = 2$, and $Q(f_3) = \max\{1,1\} = 1$. It is clear that 1 is the minimum number for which the value $Q(f_1)$ is minimum. So after the second step we will have the tree depicted in Fig 5.14.

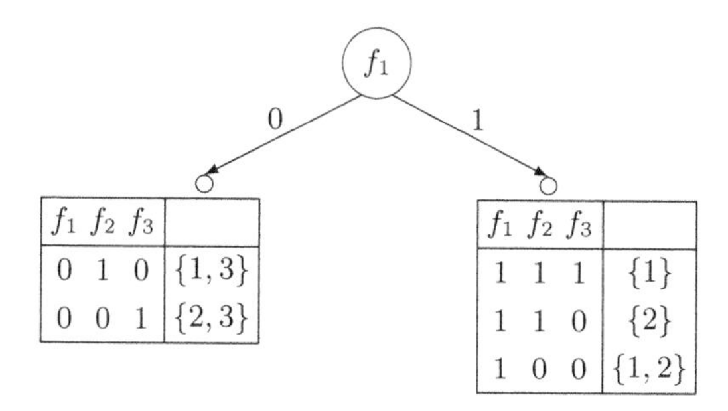


**Fig. 5.14**

We omit next steps. One can show that the result of the algorithm $U$ work for the table $T$ is the decision tree $U(T)$ represented in Fig. 5.13.

## 5.10 Conclusions

This chapter is devoted to the study of decision tables with many-valued decisions. We consider examples of such tables; relationships among decision trees, rules and tests; lower and upper bounds on the depth of decision trees, length of decision rules and cardinality of tests; approximate and exact algorithms for optimization of tests, rules and trees.

The most part of lower and upper bounds on the minimum depth of decision trees and minimum cardinality of tests considered in this chapter was published in [13, 45]. Greedy algorithm for decision tree construction similar to the algorithm considered in this chapter was studied in [51].

For $k$-valued tables filled by numbers from the set $\{0, 1, \ldots, k-1\}$, $k \geq 3$, all results considered in the chapter are true with the exception of Theorems 5.11, 5.12 and Corollary 5.19.

Instead of bounds $h(T) \geq \log_2 S(T)$ and $h(T) \geq \log_2(R(T) + 1)$, for $k$-valued tables we have bounds $h(T) \geq \log_k S(T)$ and $h(T) \geq \log_k((k-1)R(T) + 1)$. Instead of the bounds

$$\max\{M(T), \log_2 N(T)\} \leq h(T) \leq M(T) \log_2 N(T) ,$$

which are true for diagnostic 2-valued decision tables, for diagnostic $k$-valued decision tables we have bounds

$$\max\{M(T), \log_k N(T)\} \leq h(T) \leq M(T) \log_2 N(T) .$$

# 6

# Approximate Tests, Decision Trees and Rules

When we use decision trees, rules and tests as ways for knowledge representation, we would like to have relatively simple trees, rules and tests. If exact decision rules, trees or tests have large complexity, we can consider approximate trees, rules and tests.

If we use tests, decision rules or trees in classifiers, then exact tests, rules and trees can be overfitted, i.e., dependent essentially on the noise or adjusted too much to the existing examples. In this case, it is more appropriate to work with approximate tests, rules and trees.

Therefore approximate reducts [83, 88], approximate decision rules [59, 67], and approximate decision trees [8, 52, 71] are studied intensively during many years.

This chapter is devoted to the consideration of $\alpha$-tests, $\alpha$-decision trees and $\alpha$-decision rules which are special types of approximate tests, trees and rules. It contains nine sections. In Sect. 6.1, main notions are discussed. In Sect. 6.2, relationships among $\alpha$-trees, $\alpha$-rules and $\alpha$-tests are studied. In Sects. 6.3 and 6.4, lower and upper bounds on complexity of $\alpha$-rules, $\alpha$-trees and $\alpha$-tests are considered. Sections 6.5, 6.6 and 6.7 are devoted to the discussion of approximate algorithms for optimization of $\alpha$-rules, $\alpha$-trees and $\alpha$-tests. In Sect. 6.8, exact algorithms for optimization of $\alpha$-decision trees and rules are considered. Section 6.9 contains conclusions.

## 6.1 Main Notions

We will consider in this chapter only decision tables with one-valued decisions, which are filled by numbers from $\{0, 1\}$. Let $T$ be a decision table with one-valued decisions, and $T$ have $n$ columns labeled with attributes $f_1, \ldots, f_n$. A decision which is attached to the maximum number of rows in $T$ is called the *most common decision for* $T$. If we have more than one such decisions we choose the minimum one. If $T$ is empty then 1 is the most common decision for $T$. Let $\alpha$ be a real number such that $0 \leq \alpha < 1$. We define the notion of $\alpha$-decision tree for $T$.

M. Moshkov and B. Zielosko: Combinatorial Machine Learning, SCI 360, pp. 87–109.
springerlink.com © Springer-Verlag Berlin Heidelberg 2011

Let $\Gamma$ be a decision tree over $T$ and $v$ be a terminal node of $\Gamma$. Let nodes and edges in the path from the root to $v$ be labeled with attributes $f_{i_1}, \ldots, f_{i_m}$ and numbers $\delta_1, \ldots, \delta_m$ respectively. We denote by $T(v)$ the subtable $T(f_{i_1}, \delta_1) \ldots (f_{i_m}, \delta_m)$ of the table $T$. We will say that $\Gamma$ is an *$\alpha$-decision tree for $T$* if for any terminal node $v$ of $\Gamma$ the inequality $P(T(v)) \leq \alpha P(T)$ holds, $v$ is labeled with the most common decision for $T(v)$ and for any row $r$ of $T$ there exists a terminal node $v$ of $\Gamma$ such that $r$ belongs to the table $T(v)$.

We denote by $h_\alpha(T)$ the minimum depth of an $\alpha$-decision tree for $T$. It is clear that the notion of 0-decision tree for $T$ coincides with the notion of decision tree for $T$. So, $h_0(T) = h(T)$. Let $\alpha$, $\beta$ be real numbers such that $0 \leq \alpha \leq \beta < 1$. It is not difficult to show that each $\alpha$-decision tree for $T$ is also a $\beta$-decision tree for $T$. Thus, $h_\alpha(T) \geq h_\beta(T)$.

Let us define the notion of $\alpha$-test for the table $T$. An *$\alpha$-test for the table $T$* is a subset of columns $\{f_{i_1}, \ldots, f_{i_m}\}$ such that $P(T(f_{i_1}, \delta_1) \ldots (f_{i_m}, \delta_m)) \leq \alpha P(T)$ for any numbers $\delta_1, \ldots, \delta_m \in \{0, 1\}$. Empty set is an $\alpha$-test for $T$ iff $T$ is a degenerate table. An *$\alpha$-reduct for the table $T$* is an $\alpha$-test for $T$ for which each proper subset is not an $\alpha$-test. We denote by $R_\alpha(T)$ the minimum cardinality of an $\alpha$-test for the table $T$. It is clear that each $\alpha$-test has an $\alpha$-reduct as a subset. Therefore $R_\alpha(T)$ is the minimum cardinality of an $\alpha$-reduct. It is clear also that the set of tests for the table $T$ coincides with the set of 0-tests for $T$. Therefore $R_0(T) = R(T)$. Let $\alpha$, $\beta$ be real numbers such that $0 \leq \alpha \leq \beta < 1$. One can show that each $\alpha$-test for $T$ is also a $\beta$-test for $T$. Thus $R_\alpha(T) \geq R_\beta(T)$.

Let $r = (\delta_1, \ldots, \delta_n)$ be a row of $T$. A decision rule over $T$

$$f_{i_1} = b_1 \wedge \ldots \wedge f_{i_m} = b_m \rightarrow d$$

is called an *$\alpha$-decision rule for $T$ and $r$* if $b_1 = \delta_{i_1}, \ldots, b_m = \delta_{i_m}$, $d$ is the most common decision for the table $T' = T(f_{i_1}, b_1) \ldots (f_{i_m}, b_m)$ and $P(T') \leq \alpha P(T)$. We denote by $L_\alpha(T, r)$ the minimum length of $\alpha$-decision rule for $T$ and $r$. The considered decision rule is called *realizable for $r$* if $b_1 = \delta_{i_1}, \ldots, b_m = \delta_{i_m}$. The considered rule is *$\alpha$-true for $T$* if $d$ is the most common decision for $T'$ and $P(T') \leq \alpha P(T)$. A system $S$ of decision rules over $T$ is called an *$\alpha$-complete system of decision rules for $T$*, if each rule from $S$ is an $\alpha$-true for $T$ and for each row $r$ of $T$ there exists a rule from $S$ which is realizable for $r$. We denote $L(S)$ the maximum length of a rule from $S$ and by $L_\alpha(T)$ we denote the minimum value of $L(S)$ where minimum is considered among all $\alpha$-complete systems of decision rules for $T$.

*Example 6.1.* Let us consider a decision table $T$ depicted in Fig. 6.1. This is the decision table from the example Three Cups and Small Ball in Sect. 1.3.1. For this table, $P(T) = 3$. Let $\alpha$ be a real number such that $0 \leq \alpha < 1$. We consider two cases.

| $f_1$ | $f_2$ | $f_3$ | |
|---|---|---|---|
| 1 | 0 | 0 | 1 |
| 0 | 1 | 0 | 2 |
| 0 | 0 | 1 | 3 |

**Fig. 6.1**

Let $0 \leq \alpha < 1/3$. In this case, the decision tree depicted in Fig. 6.2 is an $\alpha$-decision tree for $T$ which has minimum depth. So, $h_\alpha(T) = 2$. All $\alpha$-tests for $T$ are represented in Fig. 6.3. Therefore $R_\alpha(T) = 2$. A system $S = \{f_1 = 1 \rightarrow 1, f_2 = 1 \rightarrow 2, f_3 = 1 \rightarrow 3\}$ is an $\alpha$-complete system of decision rules for $T$. Therefore $L_\alpha(T) = 1$.

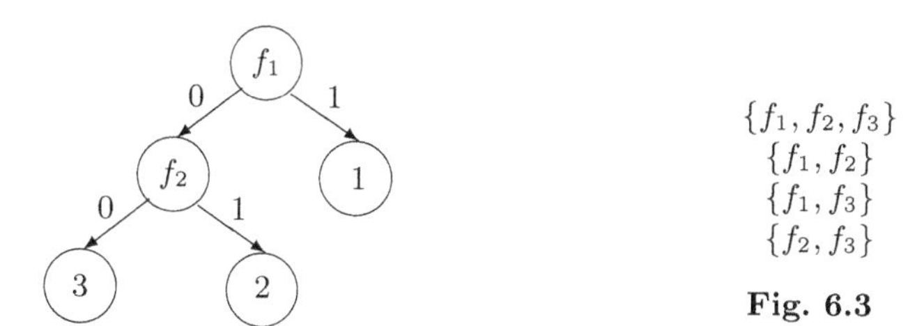


**Fig. 6.2**

**Fig. 6.3**

Let $1/3 \leq \alpha < 1$. In this case, the decision tree depicted in Fig. 6.4 is an $\alpha$-decision tree for $T$ which has minimum depth. Thus, $h_\alpha(T) = 1$. All $\alpha$-tests for $T$ are represented in Fig. 6.5. So, we have $R_\alpha(T) = 1$. A system $S = \{f_2 = 0 \rightarrow 1, f_2 = 1 \rightarrow 2\}$ is an $\alpha$-complete system of decision rules for $T$. Therefore, $L_\alpha(T) = 1$.

**Fig. 6.4**

**Fig. 6.5**

## 6.2 Relationships among $\alpha$-Trees, $\alpha$-Rules and $\alpha$-Tests

**Theorem 6.2.** *Let $T$ be a decision table with $n$ columns labeled with attributes $f_1, \ldots, f_n$ and $\alpha$ be a real number such that $0 \leq \alpha < 1$.*

1. *If $\Gamma$ is an $\alpha$-decision tree for $T$ then the set of attributes attached to working nodes of $\Gamma$ is an $\alpha$-test for the table $T$.*

2. *Let $\{f_{i_1}, \ldots, f_{i_m}\}$ be an $\alpha$-test for $T$. Then there exists an $\alpha$-decision tree $\Gamma$ for $T$ which uses only attributes from $\{f_{i_1}, \ldots, f_{i_m}\}$ and for which $h(\Gamma) = m$.*

*Proof.* 1. Let $\Gamma$ be an $\alpha$-decision tree for the table $T$. Let, for simplicity, $\{f_1, \ldots, f_t\}$ be the set of attributes attached to working nodes of $\Gamma$. Let $(\delta_1, \ldots, \delta_t) \in \{0,1\}^t$. We show that for the subtable $T' = T(f_1, \delta_1) \ldots (f_t, \delta_t)$ the inequality $P(T') \leq \alpha P(T)$ holds. Let us consider a path in $\Gamma$ from the root to a terminal node $v$ which satisfies the following condition. Let $f_{i_1}, \ldots, f_{i_m}$ be attributes attached to working nodes of this path. Then the edges of this path are labeled with numbers $\delta_{i_1}, \ldots, \delta_{i_m}$ respectively. Consider the table $T'' = T(f_{i_1}, \delta_{i_1}) \ldots (f_{i_m}, \delta_{i_m})$. It is clear that $T'' = T(v)$. Since $\Gamma$ is an $\alpha$-decision tree for $T$, we have $P(T(v)) \leq \alpha P(T)$. It is clear also that $T'$ is a subtable of $T''$. Therefore $P(T') \leq \alpha P(T)$. Taking into account that $(\delta_1, \ldots, \delta_t)$ is an arbitrary tuple from $\{0,1\}^t$ we obtain $\{f_1, \ldots, f_t\}$ is an $\alpha$-test for the table $T$.

2. Let $\{f_{i_1}, \ldots, f_{i_m}\}$ be an $\alpha$-test for the table $T$. Let us consider a decision tree $\Gamma$ over $T$ which consists of $m+1$ layers. For $j = 1, \ldots, m$, all nodes of the $j$-th layer are working nodes labeled with the attribute $f_{i_j}$. All nodes from the $(m+1)$-th layer are terminal nodes. Let $v$ be an arbitrary terminal node of $\Gamma$, and let the edges in the path from the root to $v$ be labeled with numbers $\delta_1, \ldots, \delta_m$. Then $T(v) = T(f_{i_1}, \delta_1) \ldots (f_{i_m}, \delta_m)$ and the node $v$ is labeled with the most common decision for $T(v)$. Since $\{f_{i_1}, \ldots, f_{i_m}\}$ is an $\alpha$-test for $T$, we have $P(T(v)) \leq \alpha P(T)$. Taking into account that $v$ is an arbitrary terminal node of $\Gamma$, we obtain $\Gamma$ is an $\alpha$-decision tree for $T$ for which $h(\Gamma) = m$. □

**Corollary 6.3.** *Let $T$ be a decision table, and $\alpha$ be a real number such that $0 \leq \alpha < 1$. Then*

$$h_\alpha(T) \leq R_\alpha(T) .$$

**Theorem 6.4.** *Let $T$ be a decision table with $n$ columns labeled with attributes $f_1, \ldots, f_n$, and $\alpha$ be a real number such that $0 \leq \alpha < 1$.*

1. *If $S$ is an $\alpha$-complete system of decision rules for $T$ then the set of attributes from rules in $S$ is an $\alpha$-test for $T$.*
2. *If $F = \{f_{i_1}, \ldots, f_{i_m}\}$ is a test for $T$ then there exists an $\alpha$-complete system $S$ of decision rules for $T$ which uses only attributes from $F$ and for which $L(S) = m$.*

*Proof.* 1. Let $S$ be an $\alpha$-complete system of decision rules for $T$. Let, for simplicity, $\{f_1, \ldots, f_t\}$ be the set of attributes from rules in $S$ and $(\delta_1, \ldots, \delta_t) \in \{0,1\}^t$. We show that for the table $T' = T(f_1, \delta_1) \ldots (f_t, \delta_t)$ the inequality $P(T') \leq \alpha P(T)$ holds. If $T'$ has no rows then the considered inequality is true. Let $T'$ have at least one row $\bar{\delta} = (\delta_1, \ldots, \delta_t, \delta_{t+1}, \ldots, \delta_n)$. Since $S$ is an $\alpha$-complete system of decision rules for $T$, there is a rule

$$f_{i_1} = \delta_{i_1} \wedge \ldots \wedge f_{i_m} = \delta_{i_m} \to d$$

in $S$ which is realizable for $\bar{\delta}$ and $\alpha$-true for $T$. Consider the table $T'' = T(f_{i_1}, \delta_{i_1}) \dots (f_{i_m}, \delta_{i_m})$. Since the considered rule is $\alpha$-true, $P(T'') \leq \alpha P(T)$. It is clear that $T'$ is a subtable of $T''$. Therefore $P(T') \leq \alpha P(T)$. Taking into account that $(\delta_1, \dots, \delta_t)$ is an arbitrary tuple from $\{0,1\}^t$ we obtain $\{f_1, \dots, f_t\}$ is an $\alpha$-test for the table $T$.

2. Let $F = \{f_{i_1}, \dots, f_{i_m}\}$ be an $\alpha$-test for the table $T$. For each $\bar{\delta} = (\delta_1, \dots, \delta_m) \in \{0,1\}^m$ such that the subtable

$$T(\bar{\delta}) = T(f_{i_1}, \delta_1) \dots (f_{i_m}, \delta_m)$$

is nonempty, we construct a decision rule

$$f_{i_1} = \delta_1 \wedge \dots \wedge f_{i_m} = \delta_m \to d \,,$$

where $d$ is the most common decision for $T(\bar{\delta})$. Since $F$ is an $\alpha$-test for $T$, the considered decision rule is $\alpha$-true for $T$. We denote by $S$ the set of constructed rules. It is clear that for each row of $T$ there exists a rule from $S$ which is realizable for the considered row. So, $S$ is an $\alpha$-complete system of decision rules for $T$, and $L(S) = m$. □

**Corollary 6.5.** $L_\alpha(T) \leq R_\alpha(T)$.

**Theorem 6.6.** *Let $\Gamma$ be an $\alpha$-decision tree for a decision table $T$, $0 \leq \alpha < 1$, and $S$ be the set of decision rules corresponding to paths in $\Gamma$ from the root to terminal nodes. Then $S$ is an $\alpha$-complete system of decision rules for $T$ and $L(S) = h(\Gamma)$.*

*Proof.* Since $\Gamma$ is an $\alpha$-decision tree for $T$, for each row $r$ of $T$ there exists a path $\tau$ from the root to a terminal node $v$ of $\Gamma$ such that $r$ belongs to $T(v)$, and $v$ is labeled with the most common decision for $T(v)$. It is clear that the rule rule($\tau$) corresponding to the path $\tau$ is realizable for $r$. Since $\Gamma$ is an $\alpha$-decision tree for $T$, we have $P(T(v)) \leq \alpha P(T)$. Therefore, rule($\tau$) is $\alpha$-true for $T$. It is clear that the length of rule($\tau$) is equal to the length of path $\tau$. Therefore $S$ is an $\alpha$-complete decision rule system for $T$ and $L(S) = h(\Gamma)$. □

**Corollary 6.7.** $L_\alpha(T) \leq h_\alpha(T)$.

## 6.3 Lower Bounds

We try to generalize results obtained for exact test, decision rules and decision trees to the case of approximate tests, rules and trees. Unfortunately, sometimes it is impossible. Let us show that we can not obtain a nontrivial lower bound on the value $h_\alpha(T)$ depending on the value $D(T)$ which is the number of different decisions attached to rows of the table $T$.

**Theorem 6.8.** *For any real number $\alpha$, $0 < \alpha < 1$, and for any natural $m$, there exists a decision table $T$ such that $D(T) = m + 1$ and $h_\alpha(T) = 1$.*

*Proof.* Set $r = \lceil m/2\alpha \rceil$ and consider a decision table $T$ with $r+m$ rows and $r+m$ columns. The columns of $T$ are labeled with attributes $f_1, \ldots, f_{r+m}$. The first column has 0 at the intersection with the first $r$ rows and 1 at the intersection with the last $m$ rows. For $i = 2, \ldots, r+m$, the column $f_i$ has 1 at the intersection with $i$-th row, and 0 at the intersection with all other rows. The first $r$ rows are labeled with the decision 1. The last $m$ rows are labeled with decisions $2, \ldots, m+1$ respectively (see Fig. 6.6).

| | $f_1\ f_2\ \ldots\ f_r\ \ldots\ f_{r+m}$ | |
|---|---|---|
| | 0 0 0 0 0 | 1 |
| | 0 1 0 0 0 | 1 |
| $r$ | … | … |
| | 0 0 1 0 0 | 1 |
| | 1 0 0 1 0 | 2 |
| $m$ | … | … |
| | 1 0 0 0 1 | $m+1$ |

**Fig. 6.6**

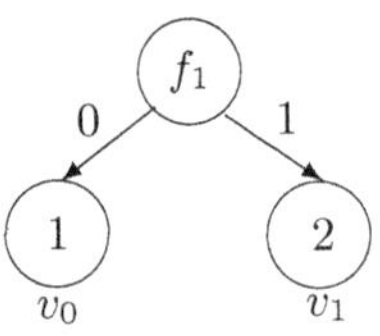

**Fig. 6.7**

It is clear that $D(T) = m+1$ and $P(T) = rm + m(m-1)/2$. Denote $T_0 = T(f_1, 0)$ and $T_1 = T(f_1, 1)$. One can easily show that $P(T_0) = 0$ and $P(T_1) = m(m-1)/2$.

Evidently, $P(T_0) < \alpha P(T)$. Let us show that $P(T_1) \leq \alpha P(T)$. Since $r = \lceil m/2\alpha \rceil$, we have $m/2\alpha \leq r$, $m/2 \leq r\alpha$, $(m-1)/2 \leq r\alpha$ and $m(m-1)/2 \leq \alpha rm$. Therefore $P(T_1) = m(m-1)/2 \leq \alpha(rm + m(m-1)/2) = \alpha P(T)$. Let us consider the decision tree depicted in Fig. 6.7. It is clear that $T(v_0) = T_0$, $T(v_1) = T_1$, 1 is the most common decision for $T_0$, and 2 is the most common decision for $T_1$. So, $\Gamma$ is an $\alpha$-decision tree for $T$. Therefore $h_\alpha(T) \leq 1$. Evidently, $P(T) > \alpha P(T)$. Hence $h_\alpha(T) > 0$ and $h_\alpha(T) = 1$. □

Another situation is with the lower bound depending on $R(T)$.

**Theorem 6.9.** *Let $T$ be a decision table and $\alpha$ be a real number such that $0 \leq \alpha < 1$. Then*

$$h_\alpha(T) \geq \log_2(R_\alpha(T) + 1) .$$

*Proof.* Let $\Gamma$ be an $\alpha$-decision tree for $T$ such that $h(\Gamma) = h_\alpha(T)$. Denote by $L_w(\Gamma)$ the number of working nodes in $\Gamma$. From Theorem 6.2 it follows that the set of attributes attached to working nodes of $\Gamma$ is an $\alpha$-test for $T$. Therefore $L_w(\Gamma) \geq R_\alpha(T)$. One can show that $L_w(\Gamma) \leq 2^{h(\Gamma)} - 1$. Hence $2^{h(\Gamma)} - 1 \geq R_\alpha(T)$, $2^{h(\Gamma)} \geq R_\alpha(T) + 1$ and $h(\Gamma) \geq \log_2(R_\alpha(T) + 1)$. Since $h(\Gamma) = h_\alpha(T)$ we obtain $h_\alpha(T) \geq \log_2(R_\alpha(T) + 1)$. □

*Example 6.10.* Consider the decision table $T$ depicted in Fig. 3.1. For this table, $P(T) = 8$.

One can show that $T$ has exactly three 1/8-reducts: $\{f_1, f_2\}$, $\{f_1, f_3\}$ and $\{f_2, f_3\}$. Therefore $R_{1/8}(T) = 2$. Using Theorem 6.9 we obtain $h_{1/8}(T) \geq$

$\log_2 3$ and $h_{1/8}(T) \geq 2$. We know that $h_0(T) = 2$ (see Example 3.3). Therefore $h_{1/8}(T) = 2$.

One can show that the table $T$ has exactly two 1/4-reducts: $\{f_2\}$ and $\{f_3\}$. Therefore $R_{1/4}(T) = 1$. Using Theorem 6.9 we obtain $h_{1/4}(T) \geq 1$. Consider the decision tree $\Gamma$ depicted in Fig. 6.8. One can show that $P(T(v_0)) = 0$,

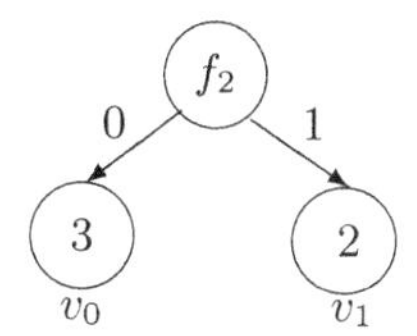


**Fig. 6.8**

$P(T(v_1)) = 2$, 3 is the most common decision for $T(v_0)$, and 2 is the most common decision for $T(v_1)$. Therefore $\Gamma$ is a 1/4-decision tree for $T$. Thus, $h_{1/4}(T) = 1$.

Let $T$ be a decision table with $n$ columns labeled with attributes $f_1, \ldots, f_n$, and $\alpha$ be a real number such that $0 \leq \alpha < 1$. Define a parameter $M_\alpha(T)$ for the table $T$. If $T$ is a degenerate table then $M_\alpha(T) = 0$. Let $T$ be a nondegenerate table, and $\bar{\delta} = (\delta_1, \ldots, \delta_n) \in \{0,1\}^n$. We denote by $M_\alpha(T, \bar{\delta})$ the minimum natural $m$ such that there exist $f_{i_1}, \ldots, f_{i_m} \in \{f_1, \ldots, f_n\}$ for which $P(T(f_{i_1}, \delta_{i_1}) \ldots (f_{i_m}, \delta_{i_m})) \leq \alpha P(T)$. Then $M_\alpha(T) = \max\{M_\alpha(T, \bar{\delta}) : \bar{\delta} \in \{0,1\}^n\}$.

Let $\alpha$ and $\beta$ be real numbers such that $0 \leq \alpha < \beta < 1$. Then $M_\alpha(T) \geq M_\beta(T)$. It is clear that $M_0(T) = M(T)$.

**Theorem 6.11.** *Let $T$ be a decision table and $\alpha$ be a real number such that $0 \leq \alpha < 1$. Then*

$$h_\alpha(T) \geq M_\alpha(T) .$$

*Proof.* If $T$ is a degenerate table then $h_\alpha(T) = 0$ and $M_\alpha(T) = 0$. Let now $T$ be a nondegenerate decision table having $n$ columns labeled with attributes $f_1, \ldots, f_n$.

Let $\Gamma$ be an $\alpha$-decision tree for $T$ such that $h(\Gamma) = h_\alpha(T)$ and $\bar{\delta} = (\delta_1, \ldots, \delta_n) \in \{0,1\}^n$ be an $n$-tuple such that $M_\alpha(T, \bar{\delta}) = M_\alpha(T)$. Let us consider a path $\tau = v_1, d_1, \ldots, v_m, d_m, v_{m+1}$ from the root $v_1$ to a terminal node $v_{m+1}$ in $\Gamma$ which satisfies the following condition: if nodes $v_1, \ldots, v_m$ are labeled with attributes $f_{i_1}, \ldots, f_{i_m}$ then edges $d_1, \ldots, d_m$ are labeled with numbers $\delta_{i_1}, \ldots, \delta_{i_m}$. We denote $T' = T(f_{i_1}, \delta_{i_1}) \ldots (f_{i_m}, \delta_{i_m})$. It is clear that $T' = T(v_{m+1})$. Since $\Gamma$ is an $\alpha$-decision tree for $T$, we have $P(T') \leq \alpha P(T)$. Therefore $m \geq M_\alpha(T, \bar{\delta})$ and $h(\Gamma) \geq M_\alpha(T, \bar{\delta})$. Since $h(\Gamma) = h_\alpha(T)$ and $M_\alpha(T, \bar{\delta}) = M_\alpha(T)$ we have $h_\alpha(T) \geq M_\alpha(T)$. $\square$

*Example 6.12.* Consider the decision table $T$ depicted in Fig. 3.3. We showed in Example 3.5 that $M_0(T) = 2$. Since $T$ is a nondegenerate table, we have

$M_\alpha(T) \geq 1$ for any $\alpha$, $0 \leq \alpha < 1$. Let us find the threshold $\beta$ such that if $\alpha < \beta$ then $M_\alpha(T) = 2$, and if $\alpha \geq \beta$ we have $M_\alpha(T) = 1$. One can show that

$$\beta = \frac{\min\{\max(P(T(f_i,0)), P(T(f_i,1)) : i = 1,2,3\}}{P(T)}$$

and $\beta = 2/8 = 0.25$.

Let $T$ be a decision table with $n$ columns labeled with attributes $f_1, \ldots, f_n$, and $m$ be a natural number such that $m \leq n$. Let us remind the notion of $(T,m)$-proof-tree.

A $(T,m)$-proof-tree is a finite directed tree $G$ with the root in which the length of each path from the root to a terminal node is equal to $m-1$. Nodes of this tree are not labeled. In each nonterminal node exactly $n$ edges start. These edges are labeled with pairs of the kind $(f_1, \delta_1) \ldots (f_n, \delta_n)$ respectively where $\delta_1, \ldots, \delta_n \in \{0,1\}$.

Let $v$ be an arbitrary terminal node of $G$ and $(f_{i_1}, \delta_1), \ldots, (f_{i_{m-1}}, \delta_{m-1})$ be pairs attached to edges in the path from the root of $G$ to the terminal node $v$. Denote $T(v) = T(f_{i_1}, \delta_1) \ldots (f_{i_{m-1}}, \delta_{m-1})$.

Let $\alpha$ be a real number such that $0 \leq \alpha < 1$. We will say that $G$ is a *proof-tree for the bound* $h_\alpha(T) \geq m$ if $P(T(v)) > \alpha P(T)$ for any terminal node $v$ of the tree $G$.

**Theorem 6.13.** *Let $T$ be a nondegenerate decision table with $n$ columns, $m$ be a natural number such that $m \leq n$, and $\alpha$ be a real number such that $0 \leq \alpha < 1$. Then a proof-tree for the bound $h_\alpha(T) \geq m$ exists if and only if the inequality $h_\alpha(T) \geq m$ holds.*

*Proof.* Let columns of $T$ be labeled with attributes $f_1, \ldots, f_n$.

1. Let $G$ be a proof-tree for the bound $h_\alpha(T) \geq m$. Let us prove that $h_\alpha(T) \geq m$. Let $\Gamma$ be an $\alpha$-decision tree for $T$ such that $h(\Gamma) = h_\alpha(T)$.
   Choose a path in $\Gamma$ from the root to some node, and a path in $G$ from the root to a terminal node in the following way. Let the root of $\Gamma$ be labeled with the attribute $f_{i_1}$. We find an edge which starts in the root of $G$ and is labeled with a pair $(f_{i_1}, \delta_1)$. We pass along this edge in the tree $G$, and pass along the edge labeled with $\delta_1$ (which starts in the root) in the tree $\Gamma$. Then we will repeat the considered procedure until we come in the tree $G$ to a terminal node $v$. In the same time, we will come to a node $w$ of the tree $\Gamma$. It is clear that $T(v)$ coincides with the subtable of $T$ consisting of rows for which during the work of $\Gamma$ we pass through the node $w$. Since $P(T(v)) > \alpha P(T)$, $w$ is not a terminal node. Therefore, the depth of $\Gamma$ is at least $m$. Since $h(\Gamma) = h_\alpha(T)$, we obtain $h_\alpha(T) \geq m$.
2. Let $h_\alpha(T) \geq m$. We prove by induction on $m$ that there exists a proof-tree for the bound $h_\alpha(T) \geq m$.
   Let $m = 1$. Then in the capacity of such proof-tree we can take the tree which consists of exactly one node. Let us assume that for some $m \geq 1$ for each decision table $T$ and for any real $\beta$, $0 \leq \beta < 1$, if $h_\beta(T) \geq m$

then there exists a proof-tree for the bound $h_\beta(T) \geq m$. Let now $T$ be a decision table, $\alpha$ be a real number such that $0 \leq \alpha < 1$, and the inequality $h_\alpha(T) \geq m+1$ hold. Let us show that there exists a proof-tree for the bound $h_\alpha(T) \geq m+1$. Let $T$ have $n$ columns labeled with $f_1, \dots, f_n$. Let $i \in \{1, \dots, n\}$. It is clear that there exists $\delta_i \in \{0,1\}$ such that $h_{\beta_i}(T(f_i, \delta_i)) \geq m$ where $\beta_i = \alpha P(T)/P(T(f_i, \delta_i))$. In the opposite case, we have $h_\alpha(T) \leq m$ which is impossible. Using inductive hypothesis we obtain that for the table $T(f_i, \delta_i)$ there exists a proof-tree $G_i$ for the bound $h_{\beta_i}(T(f_i, \delta_i)) \geq m$.

Let us construct a proof-tree $G$. In the root of $G$, $n$ edges start. These edges enter the roots of the trees $G_1, \dots, G_n$ and are labeled with pairs $(f_1, \delta_1), \dots, (f_n, \delta_n)$ respectively. One can show that $G$ is a proof-tree for the bound $h_\alpha(T) \geq m+1$. □

*Example 6.14.* It is not difficult to show that the tree depicted in Fig. 6.9 is a proof-tree for the bound $h_{0.2}(T) \geq 2$, where $T$ is the table depicted in Fig. 3.1. Using Theorem 6.13 we obtain $h_{0.2}(T) \geq 2$. Really, the tree depicted in Fig. 6.9 is a proof-tree for the bound $h_\alpha(T) \geq 2$, where $\alpha < 0.25$.

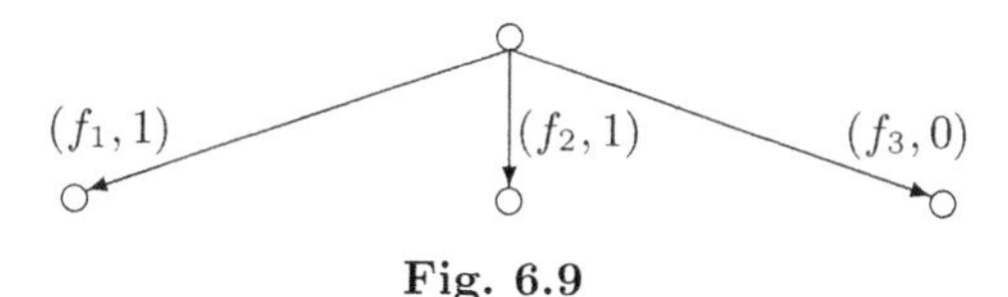


**Fig. 6.9**

**Theorem 6.15.** *Let $T$ be a decision table, $\alpha$ be a real number such that $0 \leq \alpha < 1$ and $\Delta(T)$ be the set of rows of $T$. Then $L_\alpha(T, \bar{\delta}) = M_\alpha(T, \bar{\delta})$ for any $\bar{\delta} \in \Delta(T)$ and $L_\alpha(T) = \max\{M_\alpha(T, \bar{\delta}) : \bar{\delta} \in \Delta(T)\}$.*

*Proof.* Let $T$ have $n$ columns labeled with attributes $f_1, \dots, f_n$, and $\bar{\delta} = (\delta_1, \dots, \delta_n)$ be a row of $T$. One can show that a decision rule

$$f_{i_1} = b_1 \wedge \dots \wedge f_{i_m} = b_m \rightarrow d$$

is $\alpha$-true for $T$ and realizable for $\bar{\delta}$ if and only if $b_1 = \delta_{i_1}, \dots, b_m = \delta_{i_m}$, $d$ is the most common decision for the table $T' = T(f_{i_1}, b_1) \dots (f_{i_m}, b_m)$ and $P(T') \leq \alpha P(T)$. From here it follows that $L_\alpha(T, \bar{\delta}) = M_\alpha(T, \bar{\delta})$ and $L_\alpha(T) = \max\{M_\alpha(T, \bar{\delta}) : \bar{\delta} \in \Delta(T)\}$. □

*Example 6.16.* Let us consider the decision table $T$ depicted in Fig. 3.3. In Example 3.5, we found that $\max\{M(T, \bar{\delta}) : \bar{\delta} \in \Delta(T)\} = 2$. Therefore $L_0(T) = 2$. Since $T$ is a nondegenerate table, $L_\alpha(T) \geq 1$ for any $\alpha$, $0 \leq \alpha < 1$. Let us find the threshold $\beta$ such that if $\alpha < \beta$ then $L_\alpha(T) = 2$, and if $\alpha \geq \beta$ then $L_\alpha(T) = 1$. We know (see Example 6.12) that if $\alpha \geq 0.25$ then $M_\alpha(T) = 1$ and therefore $L_\alpha(T) = 1$. Let $\alpha < 0.25$ and $\bar{\delta} = (1,1,0)$. One can show that $L_\alpha(T, \bar{\delta}) > 1$. Therefore, $L_\alpha(T) = 2$ if $\alpha < 0.25$. So $\beta = 0.25$.

## 6.4 Upper Bounds

First, we consider an upper bound of the value $R_\alpha(T)$. Let us remind that $N(T)$ is the number of rows in the table $T$.

**Theorem 6.17.** *Let $T$ be a decision table and $\alpha$ be a real number such that $0 < \alpha < 1$. Then*

$$R_\alpha(T) \leq (1 - \alpha)N(T) + 1 .$$

*Proof.* We will prove the considered inequality by induction on $N(T)$. If $N(T) = 1$ then $R_\alpha(T) = 0$ and the considered inequality holds. Let for a natural $m \geq 1$ for any decision table $T$ with $N(T) \leq m$ and for any real $\beta$, $0 < \beta < 1$, the inequality $R_\beta(T) \leq (1 - \beta)N(T) + 1$ hold.

Let $T$ be a decision table with $N(T) = m + 1$ and $\alpha$ be a real number, $0 < \alpha < 1$. If $T$ is a degenerate table then $R_\alpha(T) = 0$, and the considered inequality holds. Let us assume now that there exist two rows in $T$, which are labeled with different decisions. Let these rows be different in a column $f_i$ of the table $T$. We denote $T_0 = T(f_i, 0)$, $T_1 = T(f_i, 1)$, $N = N(T)$, $N_0 = N(T_0)$ and $N_1 = N(T_1)$. It is clear that $1 \leq N_0 \leq m$ and $1 \leq N_1 \leq m$. We consider three cases.

1. Let $P(T_0) \leq \alpha P(T)$ and $P(T_1) \leq \alpha P(T)$. In this case $\{f_i\}$ is an $\alpha$-test for the table $T$, and

$$R_\alpha(T) \leq 1 \leq (1 - \alpha)N(T) + 1 .$$

2. Let $P(T_0) \leq \alpha P(T)$ and $P(T_1) > \alpha P(T)$. We denote $\beta_1 = \alpha P(T)/P(T_1)$. It is clear that $0 < \beta_1 < 1$. Using inductive hypothesis we conclude that there exists $\beta_1$-test $B_1$ for the table $T_1$ such that $|B_1| \leq (1-\beta_1)N(T_1)+1$. It is not difficult to show that $B_1 \cup \{f_i\}$ is an $\alpha$-test for the table $T$.

   Let us prove that $\beta_1 \geq \alpha N/N_1$. To this end, we will show that $N/N_1 \leq P(T)/P(T_1)$. It is clear that $P(T) = P(T_0) + P(T_1) + P(T_0, T_1)$ where $P(T_0, T_1)$ is the number of pairs of rows $(r', r'')$ with different decisions such that $r'$ is from $T_0$ and $r''$ is from $T_1$. Thus,

$$\frac{N}{N_1} = \frac{N_1}{N_1} + \frac{N_0}{N_1} = 1 + \frac{N_0}{N_1} \text{ and } \frac{P(T)}{P(T_1)} = 1 + \frac{P(T_0)}{P(T_1)} + \frac{P(T_0, T_1)}{P(T_1)} .$$

   We will show that $N_0/N_1 \leq P(T_0, T_1)/P(T_1)$. Let $r_1, \ldots, r_{N_0}$ be all rows from $T_0$. For $i = 1, \ldots, N_0$, we denote by $P_i$ the number of pairs of rows $(r_i, r'')$ with different decisions such that $r''$ is from $T_1$. Then

$$\frac{P(T_0, T_1)}{P(T_1)} = \frac{\sum_{i=1}^{N_0} P_i}{P(T_1)} .$$

   Let us show that $P_i/P(T_1) \geq 1/N_1$ for any $i \in \{1, \ldots, N_0\}$. We consider rows of the table $T_1$. Let $b$ be the number of rows which have the same decision as $r_i$. Let $a$ be the number of rows which have other decisions. Then $P_i = a$, $P(T_1) \leq ab + a(a-1)/2$ and $N_1 = a + b$. Since $P(T_1) > \alpha P(T)$, we have $T_1$ is a nondegenerate table. Therefore, $N_1 \geq 2$ and $a \geq 1$. So,

$$\frac{P_i}{P(T_1)} \geq \frac{a}{ab + \frac{a(a-1)}{2}} = \frac{1}{b + \frac{a-1}{2}} \geq \frac{1}{b+a} .$$

Thus,

$$\frac{P(T_0, T_1)}{P(T_1)} \geq \frac{N_0}{N_1}, \quad \frac{P(T)}{P(T_1)} \geq \frac{N}{N_1}, \quad \text{and } \beta_1 = \frac{\alpha P(T)}{P(T_1)} \geq \frac{\alpha N}{N_1} .$$

Therefore,

$$\begin{aligned}|B_1 \cup \{f_1\}| &\leq (1-\beta_1)N_1 + 2 \leq \left(1 - \frac{\alpha N}{N_1}\right) N_1 + 2 \\ &= N_1 - \alpha N + 2 \leq N - \alpha N + 1 = N(1-\alpha) + 1 .\end{aligned}$$

We used here evident inequality $N_1 + 1 \leq N$.

The case $P(T_0) > \alpha P(T)$ and $P(T_1) \leq \alpha P(T)$ can be considered in the same way.

3. Let $P(T_0) > \alpha P(T)$ and $P(T_1) > \alpha P(T)$. We denote $\beta_0 = \alpha P(T)/P(T_0)$ and $\beta_1 = \alpha P(T)/P(T_1)$. It is clear that $0 < \beta_0 < 1$ and $0 < \beta_1 < 1$. Using inductive hypothesis we obtain that there exists a $\beta_0$-test $B_0$ for the table $T_0$ such that $|B_0| \leq (1-\beta_0)N_0 + 1$. Also, there exists a $\beta_1$-test $B_1$ for the table $T_1$ such that $|B_1| \leq (1-\beta_1)N_1 + 1$. It is not difficult to show that $B_0 \cup B_1 \cup \{f_i\}$ is an $\alpha$-test for the table $T$. As for the case 2, one can prove that $\beta_0 \geq \alpha N/N_0$ and $\beta_1 \geq \alpha N/N_1$. Therefore,

$$\begin{aligned}|B_0 \cup B_1 \cup \{f_i\}| &\leq \left(1 - \frac{\alpha N}{N_0}\right) N_0 + 1 + \left(1 - \frac{\alpha N}{N_1}\right) N_1 + 1 + 1 \\ &= N_0 - \alpha N + N_1 - \alpha N + 3 = N - \alpha N + 1 + 2 - \alpha N \\ &= (1-\alpha)N + 1 + 2 - \alpha N .\end{aligned}$$

Let $\alpha N \geq 2$. Then we have $R_\alpha(T) \leq (1-\alpha)N + 1$.

Let now $\alpha N < 2$. Using Theorem 3.14 we have $R_\alpha(T) \leq R_0(T) \leq N - 1 \leq N - 1 + 2 - \alpha N = (1-\alpha)N + 1$. □

We now consider an upper bound on $h_\alpha(T)$ which depends on $M_\alpha(T)$ and $\alpha$ only.

Let $T$ be a decision table with $n$ columns labeled with attributes $f_1, \ldots, f_n$, and $t$ be a nonnegative real number. We define a parameter $M^t(T)$ for the table $T$. If $T$ is a degenerate table or $t \geq P(T)$ then $M^t(T) = 0$. Let $T$ be a nondegenerate table, $t < P(T)$ and $\bar{\delta} = (\delta_1, \ldots, \delta_n) \in \{0,1\}^n$. We denote by $M^t(T, \bar{\delta})$ the minimum natural $m$ such that there exist $f_{i_1}, \ldots, f_{i_m}$ for which

$$P(T(f_{i_1}, \delta_{i_1}) \ldots (f_{i_m}, \delta_{i_m})) \leq t .$$

Then
$$M^t(T) = \max\{M^t(T, \bar{\delta}) : \bar{\delta} \in \{0,1\}^n\} .$$

**Lemma 6.18.** *Let $T$ be a decision table, $T'$ be a subtable of $T$ and $t$ be a real number such that $0 \leq t \leq P(T)$. Then*
$$M^t(T') \leq M^t(T) .$$

*Proof.* Let $\Theta$ be a decision table and $\Theta'$ be a subtable of $\Theta$. It is clear that $P(\Theta') \leq P(\Theta)$. In particular, if $\Theta$ is a degenerate table then $\Theta'$ is a degenerate table too. From here and from the definition of the parameter $M^t$ it follows that $M^t(T') \leq M^t(T)$. □

**Theorem 6.19.** *Let $T$ be a decision table and $\alpha$ be a real number such that $0 < \alpha < 1$. Then*
$$h_\alpha(T) \leq M_\alpha(T)\left(\log_2 \frac{1}{\alpha} + 1\right) .$$

*Proof.* Denote $t = \alpha P(T)$. Let $T$ be a degenerate table. Then $h_\alpha(T) = 0$, $M_\alpha(T) = 0$ and the considered inequality holds. Let $T$ be a nondegenerate table with $n$ columns labeled with attributes $f_1, \ldots, f_n$. For $i = 1, \ldots, n$, let $\sigma_i$ be a number from $\{0, 1\}$ such that
$$P(T(f_i, \sigma_i)) = \max\{P(T(f_i, 0)), P(T(f_i, 1))\} .$$

Then there exist attributes $f_{i_1}, \ldots, f_{i_m} \in \{f_1, \ldots, f_n\}$ such that $m \leq M^t(T)$ and $P(T(f_{i_1}, \sigma_{i_1}) \ldots (f_{i_m}, \sigma_{i_m})) \leq t$.

Now we begin to describe the work of an $\alpha$-decision tree $\Gamma$ on a row $r$ of the decision table $T$. First, we find sequentially values of attributes $f_{i_1}, \ldots, f_{i_m}$ on the considered row. If $f_{i_1} = \sigma_{i_1}, \ldots, f_{i_m} = \sigma_{i_m}$ then our row is localized in the subtable $T' = T(f_{i_1}, \sigma_{i_1}) \ldots (f_{i_m}, \sigma_{i_m})$ such that $P(T') \leq t$. So, the work of $\Gamma$ can be finished. The result of $\Gamma$ work on $r$ is the most common decision for the table $T'$.

Let now there exists $k \in \{1, \ldots, m\}$ such that $f_{i_1} = \sigma_{i_1}, \ldots, f_{i_{k-1}} = \sigma_{i_{k-1}}$ and $f_{i_k} \neq \sigma_{i_k}$. In this case, the considered row is localized in the subtable $T'' = T(f_{i_1}, \sigma_{i_1}) \ldots (f_{i_{k-1}}, \sigma_{i_{k-1}})(f_{i_k}, \neg\sigma_{i_k})$ where $\neg\sigma = 0$ if $\sigma = 1$ and $\neg\sigma = 1$ if $\sigma = 0$. Since $P(T(f_{i_k}, \sigma_{i_k})) \geq P(T(f_{i_k}, \neg\sigma_{i_k}))$ and $P(T) \geq P(T(f_{i_k}, \sigma_k)) + P(T(f_{i_k}, \neg\sigma_k))$, we obtain $P(T(f_{i_k}, \neg\sigma_{i_k})) \leq P(T)/2$ and $P(T'') \leq P(T)/2$.

Later the tree $\Gamma$ works similarly but instead of the table $T$ we will consider its subtable $T''$. From Lemma 6.18 it follows that $M^t(T'') \leq M^t(T)$.

The process described above will be called a big step of the decision tree $\Gamma$ work. During a big step we find values of at most $M^t(T)$ attributes. As a result, we either localize the considered row in a subtable which uncertainty is at most $t$ (and finish the work of $\Gamma$) or localize this row in a subtable which uncertainty is at most one-half of the uncertainty of initial table.

Let during the work with row $r$ the decision tree $\Gamma$ make $q$ big steps. After the big step number $q-1$ the considered row will be localized in a subtable $\Theta$ of the table $T$. Since we must make additional big step, $P(\Theta) > t = \alpha P(T)$. It is clear that $P(\Theta) \leq P(T)/2^{q-1}$. Therefore $P(T)/2^{q-1} > \alpha P(T)$ and $2^{q-1} < 1/\alpha$. Thus, $q < \log_2(1/\alpha)+1$. Taking into account that during each big step we compute values of at most $M^t(T)$ attributes and $M^t(T) = M_\alpha(T)$, we obtain $h(\Gamma) \leq M_\alpha(T)(\log_2(1/\alpha)+1)$. □

In Sect. 4.3, we considered the problem of separation of green and white points in the plane and corresponding decision table $T(S,\mu)$. From Proposition 4.28 it follows that $M(T(S,\mu)) \leq 4$. So, we have for any $\alpha$, $0 < \alpha < 1$,

**Corollary 6.20.** $h_\alpha(T(S,\mu)) \leq 4(\log_2(1/\alpha)+1)$.

Let us consider simple statement which allows us to obtain upper bounds on parameters of decision tables which are connected with the set cover problem.

**Lemma 6.21.** *Let $A$ be a finite set, and $S_1,\ldots,S_t$ be subsets of $A$ such that $S_1 \cup \ldots \cup S_t = A$. Then for any $m \in \{1,\ldots,t\}$ there exist $S_{i_1}\ldots,S_{i_m} \in \{S_1,\ldots,S_t\}$ such that $|S_{i_1} \cup \ldots \cup S_{i_m}| \geq |A|m/t$.*

*Proof.* We prove this statement by induction on $m$. Let $m=1$. Since $S_1 \cup \ldots \cup S_t = A$, there exists $S_{i_1} \in \{S_1,\ldots,S_t\}$ such that $|S_{i_1}| \geq |A|/t$. Let for some $m$, $1 \leq m < t$, the considered statement hold, i.e., there exist $S_{i_1},\ldots,S_{i_m}$ such that $|S_{i_1} \cup \ldots \cup S_{i_m}| \geq |A|m/t$.

Let us prove that the considered statement holds for $m+1$ too. Let, for the definiteness, $i_1 = 1,\ldots,i_m = m$ and $|S_1 \cup \ldots \cup S_m| = |A|m/t + x$ where $x \geq 0$. Then, evidently, there exists $S_j \in \{S_{m+1},\ldots,S_t\}$ such that

$$|S_j \setminus (S_1 \cup \ldots \cup S_m)| \geq \frac{|A| - |A|\frac{m}{t} - x}{t-m} .$$

We have that

$$\begin{aligned} |S_1 \cup \ldots \cup S_m \cup S_j| &\geq |A|\frac{m}{t} + x + \frac{|A| - |A|\frac{m}{t} - x}{t-m} \\ &= |A|\frac{m}{t} + x + |A|\frac{1}{t} - \frac{x}{t-m} \geq |A|\frac{m+1}{t} . \end{aligned}$$

□

**Proposition 6.22.** *Let $T$ be a nondegenerate decision table with $n$ columns labeled with attributes $f_1,\ldots,f_n$, $\bar{\delta} = (\delta_1,\ldots,\delta_n) \in \{0,1\}^n$, and $\alpha$ be a real number such that $0 \leq \alpha < 1$. Then*

$$M_\alpha(T,\bar{\delta}) \leq \lceil (1-\alpha)M(T,\bar{\delta}) \rceil .$$

*Proof.* Let $M(T, \bar{\delta}) = t$ and, for the definiteness, $P(T(f_1, \delta_1) \ldots (f_t, \delta_t)) = 0$. We denote by $A$ the set of all unordered pairs of rows of $T$ with different decisions. For $i = 1, \ldots, t$, we denote by $S_i$ the set of all pairs of rows from $A$ such that at least one row from the pair has in the column $f_i$ a number which is not equal to $\delta_i$. It is clear that $A = S_1 \cup \ldots \cup S_t$ and $|A| = P(T)$.

Set $m = \lceil (1 - \alpha)t \rceil$. From Lemma 6.21 it follows that there exist $S_{i_1}, \ldots, S_{i_m} \in \{S_1, \ldots, S_t\}$ such that $|S_{i_1} \cup \ldots \cup S_{i_m}| \geq |A|m/t \geq (1 - \alpha)|A|$. One can show that $P(T(f_{i_1}, \delta_{i_1}) \ldots (f_{i_m}, \delta_{i_m})) \leq \alpha P(T)$. Therefore $M_\alpha(T, \bar{\delta}) \leq \lceil (1 - \alpha) M(T, \bar{\delta}) \rceil$. □

**Corollary 6.23.** *$M_\alpha(T) \leq \lceil (1 - \alpha) M(T) \rceil$.*

**Corollary 6.24.** *$L_\alpha(T, r) \leq \lceil (1 - \alpha) L(T, r) \rceil$ for any row $r$ of $T$.*

**Corollary 6.25.** *$L_\alpha(T) \leq \lceil (1 - \alpha) L(T) \rceil$.*

**Corollary 6.26.** *$h_\alpha(T) \leq \lceil (1 - \alpha) M(T) \rceil (\log_2(1/\alpha) + 1)$.*

## 6.5 Approximate Algorithm for $\alpha$-Decision Rule Optimization

We begin from approximate algorithm for minimization of cardinality of $\alpha$-cover. Let $\alpha$ be a real number such that $0 \leq \alpha < 1$.

Let $A$ be a set containing $N > 0$ elements, and $F = \{S_1, \ldots, S_p\}$ be a family of subsets of the set $A$ such that $A = \bigcup_{i=1}^{p} S_i$. A subfamily $\{S_{i_1}, \ldots, S_{i_t}\}$ of the family $F$ will be called an *$\alpha$-cover for $A, F$* if $|\bigcup_{j=1}^{t} S_{i_j}| \geq (1 - \alpha)|A|$. The problem of searching for an $\alpha$-cover with minimum cardinality is $NP$-hard.

We consider a greedy algorithm for construction of $\alpha$-cover. During each step this algorithm chooses a subset from $F$ which covers maximum number of uncovered elements from $A$. This algorithm stops when the constructed subfamily is an $\alpha$-cover for $A, F$. We denote by $C_{\text{greedy}}(\alpha)$ the cardinality of constructed $\alpha$-cover, and by $C_{\min}(\alpha)$ we denote the minimum cardinality of $\alpha$-cover for $A, F$. The following statement was obtained by J. Cheriyan and R. Ravi in [9].

**Theorem 6.27.** *Let $0 < \alpha < 1$. Then $C_{\text{greedy}}(\alpha) < C_{\min}(0) \ln(1/\alpha) + 1$.*

*Proof.* Denote $m = C_{\min}(0)$. If $m = 1$ then, as it is not difficult to show, $C_{\text{greedy}}(\alpha) = 1$ and the considered inequality holds. Let $m \geq 2$ and $S_i$ be a subset of maximum cardinality in $F$. It is clear that $|S_i| \geq N/m$. So, after the first step we will have at most $N - N/m = N(1 - 1/m)$ uncovered elements in the set $A$. After the first step we have the following set cover problem: the set $A \setminus S_i$ and the family $\{S_1 \setminus S_i, \ldots, S_p \setminus S_i\}$. For this problem, the minimum cardinality of a cover is at most $m$. So, after the second step, when we choose a set $S_j \setminus S_i$ with maximum cardinality, the number of uncovered elements in the set $A$ will be at most $N(1 - 1/m)^2$, etc.

Let the greedy algorithm in the process of $\alpha$-cover construction make $g$ steps and construct an $\alpha$-cover of cardinality $g$. Then after the step number $g-1$ more then $\alpha N$ elements in $A$ are uncovered. Therefore $N(1-1/m)^{g-1} > \alpha N$ and $1/\alpha > (1+1/(m-1))^{g-1}$. If we take the natural logarithm of both sides of this inequality we obtain $\ln 1/\alpha > (g-1)\ln(1+1/(m-1))$. It is known that for any natural $r$, the inequality $\ln(1+1/r) > 1/(r+1)$ holds. Therefore $\ln(1/\alpha) > (g-1)/m$ and $g < m\ln(1/\alpha)+1$. Taking into account that $m = C_{\min}(0)$ and $g = C_{\text{greedy}}(\alpha)$, we obtain $C_{\text{greedy}}(\alpha) < C_{\min}(0)\ln(1/\alpha)+1$. □

We can apply the greedy algorithm for construction of $\alpha$-cover to construct $\alpha$-decision rules.

Let $T$ be a nondegenerate decision table containing $n$ columns labeled with attributes $f_1,\ldots,f_n$, $r = (b_1,\ldots,b_n)$ be a row of $T$, and $\alpha$ be a real number such that $0 < \alpha < 1$. We consider a set cover problem $A'(T,r)$, $F'(T,r) = \{S_1,\ldots,S_n\}$ where $A'(T,r)$ is the set of all unordered pairs of rows from $T$ with different decisions. For $i = 1,\ldots,n$, the set $S_i$ coincides with the set of all pairs from $A'(T,r)$ such that at least one row from the pair has at the intersection with the column $f_i$ a number different from $b_i$. One can show that the decision rule

$$f_{i_1} = b_{i_1} \wedge \ldots \wedge f_{i_m} = b_{i_m} \to d$$

is $\alpha$-true for $T$ (it is clear that this rule is realizable for $r$) if and only if $d$ is the most common decision for the table $T(f_{i_1},b_{i_1})\ldots(f_{i_m},b_{i_m})$ and $\{S_{i_1},\ldots,S_{i_m}\}$ is an $\alpha$-cover for the set cover problem $A'(T,r)$, $F'(T,r)$. Evidently, for the considered set cover problem $C_{\min}(0) = L(T,r)$.

Let us apply the greedy algorithm to the considered set cover problem. This algorithm constructs an $\alpha$-cover which corresponds to an $\alpha$-decision rule for $T$ and $r$. From Theorem 6.27 it follows that the length of this rule is at most

$$L(T,r)\ln\frac{1}{\alpha}+1\,.$$

We denote by $L_{\text{greedy}}(T,r,\alpha)$ the length of the rule constructed by the following polynomial algorithm: for a given decision table $T$, row $r$ of $T$ and $\alpha$, $0 < \alpha < 1$, we construct the set cover problem $A'(T,r)$, $F'(T,r)$ and then apply to this problem the greedy algorithm for construction of $\alpha$-cover. We transform the obtained $\alpha$-cover into an $\alpha$-decision rule for $T$ and $r$. According to what has been said above we have the following statement.

**Theorem 6.28.** *Let $T$ be a nondegenerate decision table, $r$ be a row of $T$ and $\alpha$ be a real number such that $0 < \alpha < 1$. Then*

$$L_{\text{greedy}}(T,r,\alpha) \le L(T,r)\ln\frac{1}{\alpha}+1\,.$$

*Example 6.29.* Let us apply the considered algorithm to the table $T$ depicted in Fig. 3.1, to the first row of this table and $\alpha = 1/8$.

For $i = 1,\ldots,5$, we denote by $r_i$ the $i$-th row of $T$. We have $A'(T,r_1) = \{(r_1,r_2),(r_1,r_3),(r_1,r_4),(r_1,r_5), (r_2,r_4),(r_2,r_5),(r_3,r_4),(r_3,r_5)\}$, $F'(T,r) = \{S_1,S_2,S_3\}$. $S_1 = \{(r_1,r_2),(r_1,r_4), (r_2,r_4),(r_2,r_5),(r_3,r_4)\}$, $S_2 = \{(r_1,r_4),(r_1,r_5),(r_2,r_4),(r_2,r_5),(r_3,r_4),(r_3,r_5)\}$ and $S_3 = \{(r_1,r_2),(r_1,r_3),(r_1,r_5),(r_2,r_4),(r_2,r_5),(r_3,r_4), (r_3,r_5)\}$. At the first step, the greedy algorithm chooses $S_3$. The set $\{S_3\}$ is an 1/8-cover for $A'(T,r_1)$, $F'(T,r)$. The corresponding decision rule $f_3 = 1 \rightarrow 1$ is an 1/8-decision rule for $T$ and $r_1$.

We can use the considered algorithm to construct an $\alpha$-complete decision rule system for $T$. To this end, we apply this algorithm sequentially to the table $T$, number $\alpha$ and each row $r$ of $T$. As a result, we obtain a system of rules $S$ in which each rule is $\alpha$-true for $T$ and for every row of $T$ there exists a rule from $S$ which is realizable for this row. We denote $L_{\mathrm{greedy}}(T,\alpha) = L(S)$. From Theorem 6.28 it follows

**Theorem 6.30.** *Let $T$ be a nondegenerate decision table and $\alpha$ be a real number such that $0 < \alpha < 1$. Then*

$$L_{\mathrm{greedy}}(T,\alpha) \leq L(T)\ln\frac{1}{\alpha} + 1 .$$

In Sect. 4.3, we considered the problem of separation of green and white points in the plane. From Corollary 4.29 it follows that for the table $T(S,\mu)$, corresponding to the considered problem, $L(T(S,\mu)) \leq 4$. So we have for any $\alpha$, $0 < \alpha < 1$,

**Corollary 6.31.** $L_{\mathrm{greedy}}(T(S,\mu),\alpha) \leq 4\ln(1/\alpha) + 1$.

*Example 6.32.* Let us apply the considered algorithm to the table $T$ depicted in Fig. 3.1 and $\alpha = 1/8$. As a result we obtain the following 1/8-complete decision rule system for $T$: $S = \{f_3 = 1 \rightarrow 1, f_1 = 0 \rightarrow 2, f_2 = 1 \wedge f_3 = 0 \rightarrow 2, f_2 = 0 \rightarrow 3\}$. For this system, $L(S) = 2$. One can show that $L_{1/8}(T) = 2$.

Let us consider a set cover problem $A, F$ where $A = \{a_1,\ldots,a_N\}$ and $F = \{S_1,\ldots,S_m\}$. We defined earlier the decision table $T(A,F)$. This table has $m$ columns corresponding to the sets $S_1,\ldots,S_m$ respectively, and $N+1$ rows. For $j = 1,\ldots,N$, the $j$-th row corresponds to the element $a_j$. The last $(N+1)$-th row is filled by 0. For $j = 1,\ldots,N$ and $i = 1,\ldots,m$, at the intersection of $j$-th row and $i$-th column 1 stays if and only if $a_j \in S_i$. The decision corresponding to the last row is equal to 2. All other rows are labeled with the decision 1.

One can show that a subfamily $\{S_{i_1},\ldots,S_{i_t}\}$ is an $\alpha$-cover for $A, F$, $0 \leq \alpha < 1$, if and only if the decision rule

$$f_{i_1} = 0 \wedge \ldots \wedge f_{i_t} = 0 \rightarrow d$$

is an $\alpha$-decision rule for $T(A,F)$ and the last row of $T(A,F)$ for some $d \in \{1,2\}$.

So we have a polynomial time reduction of the problem of minimization of $\alpha$-cover cardinality to the problem of minimization of $\alpha$-decision rule length. Since the first problem is $NP$-hard [83], we have

**Proposition 6.33.** *For any $\alpha$, $0 \leq \alpha < 1$, the problem of minimization of $\alpha$-decision rule length is $NP$-hard.*

Let $\alpha$ be a real number such that $0 < \alpha < 1$. Let us consider the decision table $T(A, F)$. For $j = 1, \ldots, N+1$, we denote by $r_j$ the $j$-th row of $T(A, F)$. Let $j \in \{1, \ldots, N\}$. We know that there exists a subset $S_i \in F$ such that $a_j \in S_i$. Therefore the decision rule

$$f_i = 1 \rightarrow 1$$

is an $\alpha$-decision rule for $T(A, F)$ and $r_j$. It is clear that $L_\alpha(T(A, F), r_j) \geq 1$. Hence, $L_\alpha(T(A, F), r_j) = 1$. From here it follows that $L_\alpha(T(A, F)) = L_\alpha(T(A, F), r)$ where $r = r_{N+1}$. So if we find an $\alpha$-complete decision rule system $S$ for $T(A, F)$ such that $L(S) = L_\alpha(T(A, F))$ then in this system we will find an $\alpha$-decision rule of the kind

$$f_{i_1} = 0 \wedge \ldots \wedge f_{i_t} = 0 \rightarrow d$$

for which $t = L_\alpha(T(A, F), r)$. We know that $\{S_{i_1}, \ldots, S_{i_t}\}$ is an $\alpha$-cover for $A, F$ with minimum cardinality. So we have a polynomial time reduction of the problem of minimization of $\alpha$-cover cardinality to the problem of optimization of $\alpha$-decision rule system. Since the problem of minimization of cardinality of $\alpha$-cover is $NP$-hard, we have

**Proposition 6.34.** *The problem of optimization of $\alpha$-decision rule system is $NP$-hard for any $\alpha$, $0 < \alpha < 1$.*

## 6.6 Approximate Algorithm for $\alpha$-Decision Tree Optimization

Let $\alpha$ be a real number such that $0 \leq \alpha < 1$. We now describe an algorithm $U_\alpha$ which for a given decision table $T$ constructs an $\alpha$-decision tree $U_\alpha(T)$ for the table $T$. Let $T$ have $n$ columns labeled with attributes $f_1, \ldots, f_n$. Set $s = \alpha P(T)$.

*Step* 1. Construct a tree consisting of a single node labeled with the table $T$ and proceed to the second step.

Suppose $t \geq 1$ steps have been made already. The tree obtained at the step $t$ will be denoted by $G$.

*Step* $(t+1)$. If no one node of the tree $G$ is labeled with a table then we denote by $U_\alpha(T)$ the tree $G$. The work of the algorithm $U_\alpha$ is completed.

Otherwise, we choose a node $v$ in the tree $G$ which is labeled with a subtable of the table $T$. Let the node $v$ be labeled with the table $T'$. If

$P(T') \leq s$ then instead of $T'$ we mark the node $v$ with the most common decision for $T'$ and proceed to the step $(t+2)$. Let $P(T') > s$. Then for $i = 1, \ldots, n$, we compute the value

$$Q(f_i) = \max\{P(T'(f_i, 0)), P(T'(f_i, 1))\} .$$

We mark the node $v$ with the attribute $f_{i_0}$ where $i_0$ is the minimum $i$ for which $Q(f_i)$ has the minimum value. For each $\delta \in \{0, 1\}$, we add to the tree $G$ the node $v(\delta)$, mark this node with the subtable $T'(f_{i_0}, \delta)$, draw the edge from $v$ to $v(\delta)$, and mark this edge with $\delta$. Proceed to the step $(t+2)$.

*Example 6.35.* Let us apply the algorithm $U_{0.25}$ to the table $T$ depicted in Fig. 3.1. As a result we obtain the tree $U_{0.25}$ depicted in Fig. 6.10.

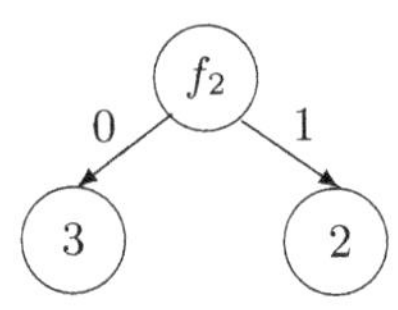


**Fig. 6.10**

We now evaluate the number of steps which the algorithm $U_\alpha$ makes during the construction of the decision tree $U_\alpha(T)$.

**Theorem 6.36.** *Let $\alpha$ be a real number such that $0 \leq \alpha < 1$, and $T$ be a decision table. Then during the construction of the tree $U_\alpha(T)$ the algorithm $U_\alpha$ makes at most $2N(T) + 1$ steps.*

The proof of this theorem is similar to the proof of Theorem 4.17. From Theorem 6.36 it follows that the algorithm $U_\alpha$ has polynomial time complexity.

**Theorem 6.37.** *Let $\alpha$ be a real number such that $0 < \alpha < 1$, and $T$ be a nondegenerate decision table. Then*

$$h(U_\alpha(T)) \leq M(T) \ln \frac{1}{\alpha} + 1 .$$

*Proof.* Let $T$ be a table with $n$ columns labeled with attributes $f_1, \ldots, f_n$. For $i = 1, \ldots, n$, we denote by $\sigma_i$ a number from $\{0, 1\}$ such that $P(T(f_i, \sigma_i)) = \max\{P(T(f_i, \sigma)) : \sigma \in \{0, 1\}\}$. It is clear that the root of the tree $U_\alpha(T)$ is labeled with attribute $f_{i_0}$ where $i_0$ is the minimum $i$ for which $P(T(f_i, \sigma_i))$ has the minimum value (it is clear that $Q(f_i) = P(T(f_i, \sigma_i))$).

As in the proof of Theorem 4.19, we can prove that

$$P(T(f_{i_0}, \sigma_{i_0})) \leq \left(1 - \frac{1}{M(T)}\right) P(T) .$$

Assume that $M(T) = 1$. From the considered inequality and from the description of algorithm $U_\alpha$ it follows that $h(U_\alpha(T)) = 1$. So if $M(T) = 1$ then the statement of theorem is true.

Let now $M(T) \geq 2$. Consider a longest path in the tree $U_\alpha(T)$ from the root to a terminal node. Let its length be equal to $k$, working nodes of this path be labeled with attributes $f_{j_1}, \ldots, f_{j_k}$, where $f_{j_1} = f_{i_0}$, and edges be labeled with numbers $\delta_1, \ldots, \delta_k$. For $t = 1, \ldots, k$, we denote by $T_t$ the table $T(f_{j_1}, \delta_1) \ldots (f_{j_t}, \delta_t)$. From Lemma 3.4 it follows that $M(T_t) \leq M(T)$ for $t = 1, \ldots, k$. We know that $P(T_1) \leq P(T)(1 - 1/M(T))$. In the same way, it is possible to prove that $P(T_t) \leq P(T)(1 - 1/M(T))^t$ for $t = 2, \ldots, k$.

Let us consider the table $T_{k-1}$. For this table, $P(T_{k-1}) \leq P(T)(1 - 1/M(T))^{k-1}$. Using the description of the algorithm $U_\alpha$ we obtain $P(T_{k-1}) > \alpha P(T)$. Therefore $\alpha < (1 - 1/M(T))^{k-1}$ and $(1 + 1/(M(T) - 1))^{k-1} < 1/\alpha$. If we take natural logarithm of both sides of this inequality we obtain $(k - 1)\ln(1 + 1/(M(T) - 1)) < \ln(1/\alpha)$. It is known that for any natural $r$ the inequality $\ln(1 + 1/r) > 1/(r + 1)$ holds. Since $M(T) \geq 2$, we obtain $(k-1)/M(T) < \ln(1/\alpha)$ and $k < M(T)\ln(1/\alpha)+1$. Taking into account that $k = h(U_\alpha(T))$ we obtain $h(U_\alpha(T)) < M(T)\ln(1/\alpha) + 1$. □

Using Theorem 3.6 we obtain

**Corollary 6.38.** *For any real $\alpha$, $0 < \alpha < 1$, and for any nondegenerate decision table $T$*

$$h(U_\alpha(T)) < h(T)\ln\frac{1}{\alpha} + 1 \,.$$

From Proposition 4.28 it follows that $M(T(S, \mu)) \leq 4$ where $T(S, \mu)$ is the decision table corresponding to the problem $S, \mu$ of separation of green and white points in the plane (see Sect. 4.3). So, for any $\alpha$, $0 < \alpha < 1$, we have

**Corollary 6.39.** $h(U_\alpha(T(S, \mu))) \leq 4\ln(1/\alpha) + 1$.

We now show that the problem of minimization of $\alpha$-decision tree depth is $NP$-hard for any $\alpha$, $0 \leq \alpha < 1$.

For a given set cover problem $A, F$, we can construct the decision table $T(A, F)$ (see previous subsection) in polynomial time. Let $A = \{a_1, \ldots, a_N\}$ and $F = \{S_1, \ldots, S_m\}$. Let $\Gamma$ be an $\alpha$-decision tree for $T(A, F)$ such that $h(\Gamma) = h_\alpha(T(A, F))$. We consider the path in $\Gamma$ in which each edge is labeled with 0. Let $\{f_{i_1}, \ldots, f_{i_t}\}$ be the set of attributes attached to working nodes of this path. One can show that $\{S_{i_1}, \ldots, S_{i_t}\}$ is an $\alpha$-cover with minimum cardinality for the problem $A, F$. So, we have a polynomial time reduction of the problem of $\alpha$-cover cardinality minimization to the problem of $\alpha$-decision tree depth minimization. Taking into account that the problem of minimization of $\alpha$-cover cardinality is $NP$-hard we obtain

**Proposition 6.40.** *For any $\alpha$, $0 \leq \alpha < 1$, the problem of minimization of $\alpha$-decision tree depth is $NP$-hard.*

## 6.7 Algorithms for $\alpha$-Test Optimization

We will prove that the problem of minimization of $\alpha$-test cardinality is $NP$-hard for any $\alpha$, $0 \leq \alpha < 1$. For a given set cover problem $A, F$, where $A = \{1, \ldots, N\}$ and $F = \{S_1, \ldots, S_m\}$, we can construct in polynomial time the decision table $T(A, F)$. One can show that a subfamily $\{S_{i_1}, \ldots, S_{i_t}\}$ is an $\alpha$-cover for $A, F$ if and only if the set of columns $\{f_{i_1}, \ldots, f_{i_t}\}$ is an $\alpha$-test for the table $T(A, F)$. So we have a polynomial time reduction of the problem of $\alpha$-cover cardinality minimization to the problem of $\alpha$-test cardinality minimization. The problem of minimization of $\alpha$-cover cardinality is $NP$-hard. As a result, we obtain

**Proposition 6.41.** *For any $\alpha$, $0 \leq \alpha < 1$, the problem of minimization of $\alpha$-test cardinality is $NP$-hard.*

Unfortunately, we do not know polynomial approximate algorithms for the problem of minimization of $\alpha$-test cardinality with nontrivial bounds of accuracy.

## 6.8 Exact Algorithms for Optimization of $\alpha$-Decision Trees and Rules

Let $\alpha$ be a real number such that $0 \leq \alpha < 1$. We describe an algorithm $W_\alpha$ which for a given nondegenerate decision table $T$ constructs an $\alpha$-decision tree for $T$ with minimum depth.

**The first part** of $W_\alpha$ work coincides with the first part of the algorithm $W$ work (see Sect. 4.3). During this part, the algorithm $W$ (and also the algorithm $W_\alpha$) constructs the set $SEP(T)$ of separable subtables of $T$ including $T$.

**The second part** of the algorithm $W_\alpha$ work is the construction of an optimal $\alpha$-decision tree $W_\alpha(T)$ for the table $T$. Begining with the smallest subtables from $SEP(T)$, the algorithm $W_\alpha$ at each step will correspond to a subtable from $SEP(T)$ a decision tree over this subtable.

Suppose that $p \geq 0$ steps of the second part of algorithm $W_\alpha$ have been made already.

*Step* $(p + 1)$: If the table $T$ in the set $SEP(T)$ is labeled with a decision tree then this tree is the result of the algorithm $W_\alpha$ work (we denote this tree by $W_\alpha(T)$). Otherwise, choose in the set $SEP(T)$ a table $D$ satisfying the following conditions:

a) the table $D$ is not labeled with a decision tree;
b) either $P(D) \leq \alpha P(T)$ or $P(D) > \alpha P(T)$ and all separable subtables of $D$ of the kind $D(f_i, \delta)$, $f_i \in E(D)$, $\delta \in \{0, 1\}$, are labeled with decision trees.

Let $P(D) \leq \alpha P(T)$ and $d$ be the most common decision for $T$. Then we mark the table $D$ with the decision tree consisting of one node which is labeled with $d$.

Otherwise, for each $f_i \in E(D)$ we construct a decision tree $\Gamma(f_i)$. The root of this tree is labeled with the attribute $f_i$. The root is the initial node of exactly two edges which are labeled with 0 and 1 respectively. These edges enter to roots of decision trees $\Gamma(f_i, 0)$ and $\Gamma(f_i, 1)$ respectively where $\Gamma(f_i, 0)$ and $\Gamma(f_i, 1)$ are decision trees attached to tables $D(f_i, 0)$ and $D(f_i, 1)$. Mark the table $D$ with one of the trees $\Gamma(f_i)$, $f_i \in E(D)$, having minimum depth, and proceed to the step $(p + 2)$.

It is not difficult to prove that after the finish of the algorithm $W_\alpha$ work each degenerate table $D$ from $SEP(T)$ will be labeled with a 0-decision tree for $D$ with minimum depth, and each nondegenerate table $D$ from $SEP(T)$ will be labeled with an $\alpha P(T)/P(D)$-decision tree for $D$ with minimum depth. Using this fact one can prove

**Theorem 6.42.** *Let $\alpha$ be a real number such that $0 \le \alpha < 1$. Then for any nondegenerate decision table $T$ the algorithm $W_\alpha$ constructs an $\alpha$-decision tree $W_\alpha(T)$ for $T$ such that $h(W_\alpha(T)) = h_\alpha(T)$, and makes exactly $2|SEP(T)| + 3$ steps. The time of the algorithm $W_\alpha$ work is bounded from below by $|SEP(T)|$, and bounded from above by a polynomial on $|SEP(T)|$ and on the number of columns in the table $T$.*

We now describe an algorithm $V_\alpha$ for the minimization of length of $\alpha$-decision rules.

**The first part** of $V_\alpha$ work coincides with the first part of the algorithm $W$ work. As a result, the set $SEP(T)$ of separable subtables of the table $T$ will be constructed.

**The second part** of the algorithm $V_\alpha$ is the construction for each row $r$ of a given decision table $T$ an $\alpha$-decision rule for $T$ and $r$ which has minimum length. Beginning with the smallest subtables from $SEP(T)$, the algorithm $V_\alpha$ at each step will correspond to each row $r$ of a subtable $T' \in SEP(T)$ a decision rule.

Suppose $p \ge 0$ steps of the second part of the algorithm $V_\alpha$ have been made already.

*Step* $(p + 1)$: If each row $r$ of the table $T$ is labeled with a decision rule then the rule attached to $r$ is the result of the work of $V_\alpha$ for $T$ and $r$ (we denote this rule by $V_\alpha(T, r)$). Otherwise, choose in the set $SEP(T)$ a table $D$ satisfying the following conditions:

a) rows of $D$ are not labeled with decision rules;
b) either $P(D) \le \alpha P(T)$, or $P(D) > \alpha P(T)$ and for all separable subtables of $D$ of the kind $D(f_i, \delta)$, $f_i \in E(D)$, $\delta \in \{0, 1\}$, each row is labeled with a decision rule.

Let $P(D) \le \alpha P(T)$ and $d$ be the most common decision for $D$. Then we mark each row of $D$ with the decision rule $\to d$.

Let $P(D) > \alpha P(T)$ and $r = (\delta_1, \ldots, \delta_n)$ be a row of $D$. For any $f_i \in E(D)$ we construct a rule rule$(r, f_i)$. Let the row $r$ in the table $D(f_i, \delta_i)$ be labeled with the rule $\beta_i \to d_i$. Then the rule rule$(r, f_i)$ is equal to $f_i = \delta_i \wedge \beta_i \to d_i$.

We mark the row $r$ of the table $D$ with one of the rules rule$(r, f_i)$, $f_i \in E(D)$, having minimum length and proceed to the step $(p+2)$.

It is not difficult to show that after the finish of the algorithm $W_\alpha$ work for each degenerate table $D \in SEP(T)$, each row $r$ of $D$ will be labeled with a 0-decision rule for $T$ and $r$ with minimum length. For each nondegenerate table $D \in SEP(T)$, each row $r$ of $D$ will be labeled with an $\alpha P(T)/P(D)$-decision rule for $T$ and $r$ having minimum length.

Using this fact one can prove

**Theorem 6.43.** *Let $\alpha$ be a real number such that $0 \leq \alpha < 1$. Then for any nondegenerate decision table $T$ and any row $r$ of $T$ the algorithm $V_\alpha$ constructs an $\alpha$-decision rule $V_\alpha(T, r)$ for $T$ and $r$ having minimum length $L_\alpha(T, r)$. During the construction of optimal rules for rows of $T$ the algorithm $V_\alpha$ makes exactly $2|SEP(T)| + 3$ steps. The time of the algorithm $V_\alpha$ work is bounded from below by $|SEP(T)|$, and bounded from above by a polynomial on $|SEP(T)|$ and on the number of columns in the table $T$.*

## 6.9 Conclusions

This chapter is devoted to the study of $\alpha$-tests, $\alpha$-decision trees and $\alpha$-decision rules. We consider relationships among these objects, bounds on complexity and algorithms for construction of such trees, rules and tests.

The bound from Theorem 6.17 and a statement close to Theorem 6.19 were published in [14, 52]. An algorithm similar to the algorithm $W_\alpha$ for optimization of $\alpha$-decision trees was considered in [2].

Note that there are different approaches to the definition of notions of approximate decision trees, rules and tests. In particular, in the book [59] $\alpha$-tests and $\alpha$-decision rules are studied which are defined in other ways than in this chapter.

Let $T$ be a decision table with $n$ columns labeled with attributes $f_1, \ldots, f_n$. A subset $B$ of the set $\{f_1, \ldots, f_n\}$ is called an $\alpha$-test for $T$ if attributes from $B$ separate at least $(1-\alpha)P(T)$ unordered pairs of rows with different decisions from $T$ (an attribute $f_i$ separates two rows if these rows have different numbers at the intersection with the column $f_i$).

Let $r = (\delta_1, \ldots, \delta_n)$ be a row of $T$ labeled with the decision $d$. We denote by $P(T, r)$ the number of rows from $T$ with decisions different from $d$. A decision rule

$$f_{i_1} = \delta_{i_1} \wedge \ldots \wedge f_{i_m} = \delta_{i_m} \rightarrow d$$

is called an *$\alpha$-decision rule for $T$ and $r$* if attributes $f_{i_1}, \ldots, f_{i_m}$ separate from $r$ at least $(1-\alpha)P(T, r)$ rows with decisions different from $d$.

The book [59] contains bounds on complexity and algorithms for construction of such $\alpha$-tests and $\alpha$-decision rules. In contrast with this chapter, in [59] it is proven that, under some natural assumptions on the class $NP$, a simple greedy algorithm is close to the best polynomial approximate algorithms for the minimization of $\alpha$-test cardinality.

Next four chapters are devoted to the consideration of various applications of tools created in the first part of book. In Chap. 7, we discuss the use of tests, decision trees and rules in supervised machine learning including lazy learning algorithms. Chapter 8 is devoted to the study of complexity of decision trees and decision rules over infinite systems of attributes. In Chap. 9, we study decision trees with so-called quasilinear attributes, and applications of the obtained results to problems of discrete optimization and analysis of acyclic programs. In Chap. 10, we consider two more applications: the diagnosis of constant faults in combinatorial circuits and the recognition of regular language words.

# Part II

# Applications

# 7

# Supervised Learning

In the previous chapters, we considered algorithms for construction of classifiers—decision trees and decision rule systems for a given decision table $T$. If $T$ contains complete information (we know all possible tuples of values of attributes, and these tuples are rows of $T$) then depending on our aims we should construct either exact or approximate classifiers. In the last case, we can control the accuracy of approximate classifiers.

If $T$ contains incomplete information (we do not know all possible tuples of values of attributes and corresponding decisions) then we have essentially more complicated problem known as *supervised learning*. For a given decision table $T$ with conditional attributes $f_1, \ldots, f_n$ and the decision attribute $d$, we should construct a classifier which will predict values of the decision attribute for tuples of values of conditional attributes which, possible, are not rows of the table $T$. In this case, exact classifiers can be overfitted, i.e., have a good accuracy for $T$ and a bad one for tuples of values of attributes that are not rows of $T$.

The usual way in this situation is to divide initial table $T$ into three subtables: training subtable $T_1$, validation subtable $T_2$ and test subtable $T_3$. The subtable $T_1$ is used for construction of initial classifier. The subtable $T_2$ is used for pruning of this classifier: we step by step decrease the accuracy of the classifier relative to $T_1$ by removal of its parts (nodes of decision tree or conditions from the left-hand side of decision rules), and stop when the accuracy of obtained classifier relative to $T_2$ will be maximum. The subtable $T_3$ is used to evaluate the accuracy of classifier obtained after pruning. If the accuracy is enough good we can use this classifier to predict decisions for tuples of values of attributes that are not rows of $T$.

In this chapter, we consider three known approaches to the supervised learning problem: based on decision trees (see, for example, [8, 71]), based on decision rule systems (see, for example, [73]) and so-called lazy learning algorithms (we omit the construction of classifier and work directly with input tuple of attribute values and decision table $T$ [1, 20]).

M. Moshkov and B. Zielosko: Combinatorial Machine Learning, SCI 360, pp. 113–126.
springerlink.com © Springer-Verlag Berlin Heidelberg 2011

This chapter contains four sections. In Sect. 7.1, we consider classifiers based on decision trees. In Sect. 7.2, we study classifiers based on decision rules. Section 7.3 is devoted to the consideration of lazy learning algorithms. Section 7.4 contains conclusions.

## 7.1 Classifiers Based on Decision Trees

We studied two ways for exact decision tree construction: based on greedy algorithm (algorithm $U$) and based on dynamic programming approach (algorithm $W$). Modifications of these algorithms (algorithms $U_\alpha$ and $W_\alpha$) allow us to construct approximate trees—$\alpha$-decision trees, $0 \leq \alpha < 1$.

The considered algorithms are trying to minimize or minimize the depth of decision trees. We can have also other aims, for example, to minimize the number of nodes in decision trees. It is easy to modify algorithms $W$ and $W_\alpha$ for the minimization of number of nodes in exact and $\alpha$-decision trees (see [12, 2]. If in the algorithms $U$ and $U_\alpha$ under the selection of attribute we minimize the parameter $P(T(f_i, 0)) + P(T(f_i, 1))$ instead of the parameter $\max\{P(T(f_i, 0)), P(T(f_i, 1))\}$ we will obtain algorithms which are more adjusted to the minimization of the number of nodes in decision trees.

Let $\Gamma$ be an $\alpha$-decision tree for $T$. We can use $\Gamma$ as a classifier to predict the value of decision attribute for a tuple $\bar{\delta}$ of values of conditional attributes which is not a row of $T$. Let us describe the work of $\Gamma$ on $\bar{\delta}$. We begin from the root of $\Gamma$. Let us assume that we reached a node $v$ of $\Gamma$. If $v$ is a terminal node of $\Gamma$ labeled with a decision $c$, then $c$ is the result of work of $\Gamma$ on $\bar{\delta}$. Let $v$ be a nonterminal node labeled with an attribute $f_i$, and $\delta_i$ be the value of $f_i$ in the tuple $\bar{\delta}$. If there is no edge which issues from $v$ and is labeled with $\delta_i$ then the result of $\Gamma$ work on $\bar{\delta}$ is the most common decision for the table $T(v)$ (see explanation of the notation $T(v)$ in the next paragraph). Otherwise, we pass along the edge that issues from $v$ and is labeled with $\delta_i$, etc.

Let us assume now that we divided given nondegenerate decision table $T$ into three nondegenerate subtables $T_1$, $T_2$ and $T_3$, and constructed an $\alpha$-decision tree $\Gamma$ for the subtable $T_1$. We describe now the procedure of decision tree $\Gamma$ pruning based on decision tables $T_1$ and $T_2$. For each node $v$ of $\Gamma$, we construct a subtable $T_1(v)$ of the table $T_1$, where $T_1(v) = T_1$ if $v$ is the root of $\Gamma$, and $T_1(v) = T_1(f_{i_1}, a_1) \ldots (f_{i_m}, a_m)$ if $v$ is not a root of $\Gamma$, $f_{i_1}, \ldots, f_{i_m}$ are attributes attached to nodes of the path from the root of $\Gamma$ to $v$, and $a_1, \ldots, a_m$ are numbers attached to edges of this path. We denote $\alpha(v) = P(T_1(v))/P(T_1)$.

Let $\Gamma$ contain $t$ nonterminal nodes, and $v_1, \ldots, v_t$ be all nonterminal nodes of $\Gamma$ in an order such that $\alpha(v_1) \leq \alpha(v_2) \leq \ldots \leq \alpha(v_t)$, and for any $i \in \{1, \ldots, t-1\}$, if $\alpha(v_i) = \alpha(v_{i+1})$ then the distance from the root of $\Gamma$ to $v_i$ is at least the distance from the root to $v_{i+1}$. We construct now a sequence $\Gamma_0, \Gamma_1, \ldots, \Gamma_t$ of decision trees. Let $\Gamma_0 = \Gamma$, and let us assume that for some $i \in \{0, \ldots, t-1\}$ the decision tree $\Gamma_i$ is already constructed. We now construct the decision tree $\Gamma_{i+1}$. Let $D$ be a subtree of $\Gamma_i$ with the root $v_{i+1}$. We remove from $\Gamma_i$ all nodes and edges of $D$ with the exception of $v_{i+1}$. We transform the node $v_{i+1}$ into a terminal node which is labeled with the most common decision for $T_1(v_{i+1})$. As a result, we obtain the decision tree $\Gamma_{i+1}$.

For $i = 0, \ldots, t$, we apply the decision tree $\Gamma_i$ to each row of the table $T_2$ and find the number of misclassifications—the number of rows in $T_2$ for which the result of $\Gamma_i$ work does not equal to the decision attached to the considered row.

We choose minimum $i_0 \in \{0, \ldots, t\}$ for which the tree $\Gamma_{i_0}$ has the minimum number of misclassifications. This tree will be considered as the final classifier. We apply $\Gamma_{i_0}$ to the table $T_3$ and evaluate its quality—the number of misclassifications on the rows of $T_3$.

Note that we can consider different approaches to the pruning of decision trees.

*Example 7.1.* Let us consider the decision table $T$ depicted in Fig. 3.1 and the decision tree $\Gamma$ depicted in Fig. 3.2. We know that $\Gamma$ is a decision tree for $T$ with minimum depth. We can apply $\Gamma$ to the three tuples of values of conditional attributes $f_1, f_2, f_3$ which are not rows of $T$ to predict the values of decision attribute $d$ (see Fig. 7.1).

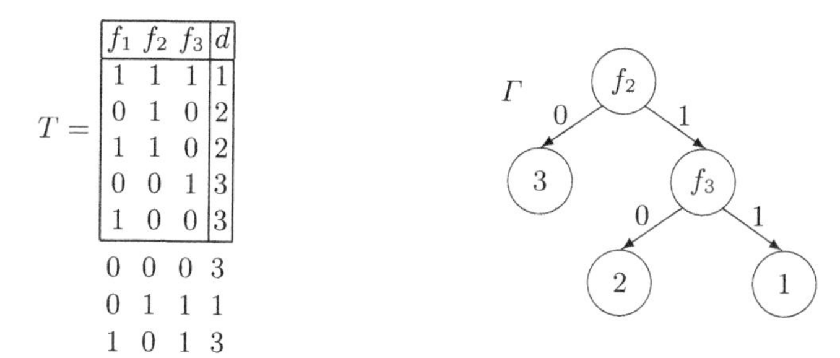


**Fig. 7.1**

## 7.2 Classifiers Based on Decision Rules

We considered a number of ways to construct for a given decision table $T$ a complete system of decision rules or an $\alpha$-complete system of decision rules, $0 \leq \alpha < 1$. First, we should list these ways.

### *7.2.1 Use of Greedy Algorithms*

We apply to each row $r$ of $T$ the greedy algorithm (see Sect. 4.1.3) which constructs a decision rule for $T$ and $r$. As a result, we obtain a complete system of decision rules for $T$.

Let $\alpha$ be a real number such that $0 \leq \alpha < 1$. We can apply to each row $r$ of $T$ the greedy algorithm (see Sect. 6.5) which constructs an $\alpha$-decision rule for $T$ and $r$. As a result, we obtain an $\alpha$-complete system of decision rules for $T$.

*Example 7.2.* Let us apply the greedy algorithm for construction of exact decision rules and the greedy algorithm for construction of 1/8-decision rules to each row of the decision table $T$ depicted in Fig. 3.1 (see also Fig. 7.1). As a result, we obtain a complete system of decision rules for $T$

$$\{f_1 = 1 \wedge f_3 = 1 \rightarrow 1, f_1 = 0 \wedge f_2 = 1 \rightarrow 2, f_2 = 1 \wedge f_3 = 0 \rightarrow 2,$$
$$f_2 = 0 \rightarrow 3, f_2 = 0 \rightarrow 3\} ,$$

and a 1/8-complete system of decision rules for $T$

$$\{f_3 = 1 \rightarrow 1, f_1 = 0 \rightarrow 2, f_2 = 1 \wedge f_3 = 0 \rightarrow 2, f_2 = 0 \rightarrow 3, f_2 = 0 \rightarrow 3\} .$$

### 7.2.2 Use of Dynamic Programming Approach

We can apply to each row $r$ of $T$ the algorithm $V$ (see Sect. 4.3) which constructs a decision rule for $T$ and $r$ with minimum length. As a result, we obtain a complete system of decision rules for $T$.

Let $0 \leq \alpha < 1$. We apply to each row $r$ of $T$ the algorithm $V_\alpha$ (see Sect. 6.8) which constructs an $\alpha$-decision rule for $T$ and $r$ with minimum length. As a result, we obtain an $\alpha$-complete system of decision rules for $T$.

### 7.2.3 From Test to Complete System of Decision Rules

Let $F = \{f_{i_1}, \ldots, f_{i_m}\}$ be a test for $T$. For each row $r$ of $T$, we construct the decision rule

$$f_{i_1} = a_1 \wedge \ldots \wedge f_{i_m} = a_m \rightarrow t$$

where $a_1, \ldots, a_m$ are numbers at the intersection of the row $r$ and columns $f_{i_1}, \ldots, f_{i_m}$ and $t$ is the decision attached to the row $r$. Since $F$ is a test for $T$, the obtained system of decision rules is a complete system for $T$ (see Theorem 2.25).

Let $\alpha$ be a real number, $0 \leq \alpha < 1$, and $F = \{f_{i_1}, \ldots, f_{i_m}\}$ be an $\alpha$-test for $T$. For each row $r$ of $T$, we construct the decision rule

$$f_{i_1} = a_1 \wedge \ldots \wedge f_{i_m} = a_m \rightarrow t$$

where $a_1, \ldots, a_m$ are numbers at the intersection of the row $r$ and columns $f_{i_1}, \ldots, f_{i_m}$ and $t$ is the most common decision for the table $T(f_{i_1}, a_1) \ldots (f_{i_m}, a_m)$. Since $F$ is an $\alpha$-test for $T$, the obtained system of decision rules is an $\alpha$-complete system for $T$ (see Theorem 6.4).

*Example 7.3.* Let us consider the decision table $T$ depicted in Fig. 3.1. We know that $\{f_2, f_3\}$ is a test for $T$. One can show that $\{f_2\}$ is a 1/4-test for $T$. Therefore

$$\{f_2 = 1 \wedge f_3 = 1 \rightarrow 1, f_2 = 1 \wedge f_3 = 0 \rightarrow 2,\ f_2 = 1 \wedge f_3 = 0 \rightarrow 2,\\ f_2 = 0 \wedge f_3 = 1 \rightarrow 3, f_2 = 0 \wedge f_3 = 0 \rightarrow 3\}$$

is a complete system of decision rules for $T$, and

$$\{f_2 = 1 \rightarrow 2, f_2 = 1 \rightarrow 2, f_2 = 1 \rightarrow 2, f_2 = 0 \rightarrow 3, f_2 = 0 \rightarrow 3\}$$

is a 1/4-complete system of decision rules for $T$.

### 7.2.4 From Decision Tree to Complete System of Decision Rules

Let $\Gamma$ be an $\alpha$-decision tree, $0 \leq \alpha < 1$. If $\alpha = 0$ then we have an exact decision tree. Let $\tau$ be a path in $\Gamma$ from the root to a terminal node in which working nodes are labeled with attributes $f_{i_1}, \ldots, f_{i_m}$, edges are labeled with numbers $b_1, \ldots, b_m$, and the terminal node of $\tau$ is labeled with the decision $t$. We correspond to $\tau$ the decision rule

$$f_{i_1} = b_1 \wedge \ldots \wedge f_{i_m} = b_m \rightarrow t\ .$$

We know that the set of decision rules corresponding to paths in $\Gamma$ from the root to terminal nodes is an $\alpha$-complete system of decision rules for $T$ (see Theorem 6.6). In particular, if $\Gamma$ is a decision tree for $T$ then the considered system of decision rules is a complete system for $T$ (see Theorem 2.27).

*Example 7.4.* Let us consider the decision table $T$ depicted in Fig. 3.1 and the decision tree $\Gamma$ for $T$ depicted in Fig. 3.2 (see also Fig. 7.1). Then

$$\{f_2 = 0 \rightarrow 3, f_2 = 1 \wedge f_3 = 0 \rightarrow 2, f_2 = 1 \wedge f_3 = 1 \rightarrow 1\}$$

is the set of decision rules corresponding to paths in $\Gamma$ from the root to terminal nodes. This system is a complete system of decision rules for $T$.

### 7.2.5 Simplification of Rule System

Let $0 \leq \alpha < 1$ and $S$ be an $\alpha$-complete system of decision rules for $T$ constructed by one of the considered algorithms. We can try to simplify $S$: try to minimize the total length of rules in $S$ and the number of rules in $S$. Let

$$f_{i_1} = a_1 \wedge \ldots \wedge f_{i_m} = a_m \rightarrow t \tag{7.1}$$

be a rule from $S$. We try to remove from the left-hand side of the rule (7.1) as much as possible conditions $f_{i_j} = a_j$, $j \in \{1, \ldots, m\}$, such that for remaining conditions $f_{i_{j(1)}} = a_{j(1)}, \ldots, f_{i_{j(k)}} = a_{j(k)}$ for subtable $T' = T(f_{i_{j(1)}}, a_{j(1)}) \ldots (f_{i_{j(k)}}, a_{j(k)})$ the inequality $P(T') \leq \alpha P(T)$ holds. Then instead of rule (7.1) we add to $S$ the rule

$$f_{i_{j(1)}} = a_{j(1)} \wedge \ldots \wedge f_{i_{j(k)}} = a_{j(k)} \rightarrow t' \tag{7.2}$$

where $t'$ is the most common decision for $T'$. It is clear that the set of rules obtained from $S$ by substitution of reduced rule (7.2) for each rule (7.1) from $S$ is an $\alpha$-decision rule system for $T$.

We will say that a rule from $S$ *covers* a row $r$ from $T$ if this rule is realizable for $r$. Since $S$ is an $\alpha$-complete system of decision rules for $T$, rules from $S$ cover all rows from $T$. Let $S'$ be a subsystem of $S$ such that rules from $S'$ cover all rows from $T$. It is clear that $S'$ is an $\alpha$-complete system of decision rules for $T$. We can try to minimize the number of rules in $S'$. To this end, we can use the greedy algorithm for set cover problem.

### 7.2.6 System of Rules as Classifier

Let $T$ be a decision table with conditional attributes $f_1, \ldots, f_n$ and the decision attribute $d$. Let $S$ be an $\alpha$-complete system of decision rules for $T$. We can use $S$ as a classifier for the prediction of value of decision attribute $d$ for a tuple $\bar{a} = (a_1, \ldots, a_n)$ of values of conditional attributes $f_1, \ldots, f_n$ in the case when $\bar{a}$ is not a row of $T$. We will say that a decision rule

$$f_{i_1} = b_1 \wedge \ldots \wedge f_{i_m} = b_m \rightarrow t$$

is *realizable for* $\bar{a}$ if $b_1 = a_{i_1}, \ldots, b_m = a_{i_m}$. If $S$ does not contain rules which are realizable for $\bar{a}$ then the value of the attribute $d$ for $\bar{a}$ will be equal to the most common decision for $T$. Let $S$ contain rules which are realizable for $\bar{a}$ and $c$ be the minimum value of $d$ such that the number of rules from $S$ which are realizable for $\bar{a}$ and have $c$ on the right-hand side is maximum. Then the value of $d$ for $\bar{a}$ will be equal to $c$.

### 7.2.7 Pruning

A complete system of rules or an $\alpha$-complete system of rules with small value of $\alpha$ can be overfitted, i.e., can have high accuracy on $T$ and low accuracy for tuples which are not rows of $T$. To improve this situation, we can use a procedure of pruning of decision rules.

Let a nondegenerate decision table $T$ is divided into three nondegenerate subtables $T_1$, $T_2$ and $T_3$. We use one of the considered approaches to construct an $\alpha$-complete system of decision rules for $T_1$. After that, we can simplify this system. Let $S$ be the obtained $\alpha$-complete system of decision rules for $T_1$. Let us consider an arbitrary decision rule from $S$

$$f_{i_1} = a_1 \wedge \ldots \wedge f_{i_m} = a_m \to t \,. \tag{7.3}$$

For $j \in \{1, \ldots, m\}$, we consider the *subrule*

$$f_{i_j} = a_j \wedge \ldots \wedge f_{i_m} = a_m \to t' \tag{7.4}$$

of the rule (7.3), where $t'$ is the most common decision for the subtable $T_1(f_{i_j}, a_j) \ldots (f_{i_m}, a_m))$, and find *inaccuracy*

$$\frac{P(T_1(f_{i_j}, a_j) \ldots (f_{i_m}, a_m))}{P(T_1)}$$

of the subrule (7.4) relative to the table $T_1$.

Let $\alpha_1 < \alpha_2 < \ldots < \alpha_q$ be all different inaccuracies for all subrules of rules from $S$ which are greater than or equal to $\alpha$. For every $k \in \{1, \ldots, q\}$, we construct a system of decision rules $S_k$. For any rule (7.3) from $S$ we add to $S_k$ a subrule (7.4) of (7.3) with maximum $j \in \{1, \ldots, m\}$ for which the inaccuracy of (7.4) is at most $\alpha_k$.

For each $k \in \{1, \ldots, q\}$, we apply the decision rule system $S_k$ as classifier to the table $T_2$ and find the number of misclassifications—the number of rows in $T_2$ for which the result of $S_k$ work does not equal to the decision attached to the considered row.

We choose minimum $k_0 \in \{1, \ldots, q\}$ for which the rule system $S_{k_0}$ has the minimum number of misclassifications. This system will be considered as the final classifier. We apply $S_{k_0}$ to the table $T_3$ and evaluate its quality—the number of misclassifications on the rows of $T_3$.

*Example 7.5.* Let us consider the decision table $T$ depicted in Fig. 3.1 and 1/8-complete system of decision rules for $T$

$$S = \{f_3 = 1 \to 1, f_1 = 0 \to 2, f_2 = 1 \wedge f_3 = 0 \to 2, f_2 = 0 \to 3\}$$

constructed by the greedy algorithm (see Example 7.2, we removed from the considered system one of two equal rules $f_2 = 0 \to 3$). We apply $S$ to three tuples of values of attributes $f_1, f_2, f_3$ which are not rows of $T$ to predict values of the decision attribute $d$ (see Fig. 7.2).

## 7.3 Lazy Learning Algorithms

Let $T$ be a decision table with $n$ conditional attributes $f_1, \ldots, f_n$ and the decision attribute $d$.

Instead of construction of classifier, we can use information contained in the decision table $T$ directly to predict the value of decision attribute for a tuple of values of conditional attributes which is not a row of $T$ [1].

$T =$

| $f_1$ | $f_2$ | $f_3$ | $d$ |
|---|---|---|---|
| 1 | 1 | 1 | 1 |
| 0 | 1 | 0 | 2 |
| 1 | 1 | 0 | 2 |
| 0 | 0 | 1 | 3 |
| 1 | 0 | 0 | 3 |

0 0 0 2
0 1 1 1
1 0 1 1

**Fig. 7.2**

### 7.3.1 k-Nearest Neighbor Algorithm

Let a distance function is defined on the set of possible tuples of attribute $f_1, \ldots, f_n$ values, $k$ be a natural number and $\bar{\delta}$ be a tuple of values of attributes $f_1, \ldots, f_n$ which is not a row of $T$. To assign a value of the decision attribute $d$ to $\bar{\delta}$ we find $k$ rows $\sigma_1, \ldots, \sigma_k$ from $T$ which are nearest to $\bar{\delta}$ (relative to the considered distance). More precisely, we put all rows of $T$ in order such that the $i$-th row $r_i$ precedes the $j$-th row $r_j$ if the distance from $\bar{\delta}$ to $r_i$ is less than the distance from $\bar{\delta}$ to $r_j$. If the considered distances are equal then $r_i$ precedes $r_j$ if $i < j$. We assign to $\bar{\delta}$ the minimum decision which is attached to the maximum number of rows $\sigma_1, \ldots, \sigma_k$ (see also [16]).

*Example 7.6.* Let us consider the decision table $T$ depicted in Fig. 3.1. We apply $3NN$ ($k = 3$) algorithm based on Hamming distance to the table $T$ and to three tuples of values of attributes $f_1$, $f_2$, $f_3$ which are not rows of $T$ to predict values of the decision attribute $d$ (see Fig. 7.3). The Hamming distance between two tuples of attribute $f_1, \ldots, f_n$ values is the number of attributes for which the considered tuples have different values. In particular, the Hamming distance between the tuple $(0, 0, 0)$ and any row of $T$ is equal to the number of units in the considered row. There are three rows for which the distance is equal to 1: $(0, 1, 0)$, $(0, 0, 1)$ and $(1, 0, 0)$. These rows are labeled with decisions 2, 3, 3. Therefore we assign the decision 3 to the tuple $(0, 0, 0)$.

### 7.3.2 Lazy Decision Trees and Rules

Let $\bar{\delta} = (\delta_1, \ldots, \delta_n)$ be a tuple of values of attributes $f_1, \ldots, f_n$ which is not a row of $T$.

Instead of construction of an $\alpha$-decision tree $\Gamma$ for $T$ by greedy algorithm and use $\Gamma$ to assign a decision to $\bar{\delta}$, we can simulate the work of $\Gamma$ on $\bar{\delta}$ by construction of corresponding path from the root of $\Gamma$ to some node (see description of the work of $\Gamma$ on $\bar{\delta}$ in Sect. 7.1).

| | $f_1$ | $f_2$ | $f_3$ | $d$ |
|---|---|---|---|---|
| $T =$ | 1 | 1 | 1 | 1 |
| | 0 | 1 | 0 | 2 |
| | 1 | 1 | 0 | 2 |
| | 0 | 0 | 1 | 3 |
| | 1 | 0 | 0 | 3 |

| | | | |
|---|---|---|---|
| 0 | 0 | 0 | 3 |
| 0 | 1 | 1 | 1 |
| 1 | 0 | 1 | 3 |

**Fig. 7.3**

Let us assume that we already constructed a path $v_1, \ldots, v_m$ from the root $v_1$ of $\Gamma$ to a node $v_m$ of $\Gamma$. Let nodes $v_1, \ldots, v_{m-1}$ be labeled with attributes $f_{i_1}, \ldots, f_{i_{m-1}}$. Set $T(v_m) = T(f_{i_1}, \delta_{i_1}) \ldots (f_{i_{m-1}}, \delta_{i_{m-1}})$. If $P(T(v_m)) \leq \alpha P(T)$ then $v_m$ is a terminal node of $\Gamma$. We assign to $v_m$ (and to $\bar{\delta}$) the most common decision for $T(v_m)$.

Let $P(T(v_m)) > \alpha P(T)$ and $i_m$ is the minimum number from $\{1, \ldots, n\}$ for which the column $f_{i_m}$ in $T(v_m)$ contains different numbers and

$$\max_{\sigma \in D_{i_m}} P(T(v_m)(f_{i_m}, \sigma)) = \min\{\max_{\sigma \in D_j} P(T(v_m)(f_j, \sigma)) : j = 1, \ldots, n\}$$

where $D_j$, $j \in \{1, \ldots, n\}$, is the set of values of the attribute $f_j$ in the table $T(v_m)$. If $T(v_m)(f_{i_m}, \delta_{i_m})$ is the empty table then we assign to $\bar{\delta}$ the most common decision for $T(v_m)$. Otherwise, we attach the attribute $f_{i_m}$ to the node $v_m$, add new node $v_{m+1}$ and the edge from $v_m$ to $v_{m+1}$ which is labeled with the number $\delta_{i_m}$, etc.

We will say about this approach to prediction as about *lazy decision trees.*

To avoid the appearance of empty table $T(v_m)(f_{i_m}, \delta_{i_m})$ as long as possible we can modify a bit the described procedure and choose $f_{i_m}$ only among such $f_j$, $j \in \{1, \ldots, n\}$, for which the table $T(v_m)(f_j, \delta_j)$ is nonempty. If there are no such $f_j$ then we assign to $\bar{\delta}$ the most common decision for $T(v_m)$.

After the considered modification, we can not say that we simulate the work of an $\alpha$-decision tree. However, we will say about this approach to prediction also as about lazy decision trees.

We can consider not only lazy decision trees but also lazy decision rules: instead of construction of an $\alpha$-complete system of decision rules for $T$ we can construct an $\alpha$-decision rule for $T$ and $\bar{\delta}$ by slightly modified greedy algorithm.

Let us assume that we already constructed a prefix $f_{i_1} = \delta_{i_1} \wedge \ldots \wedge f_{i_{m-1}} = \delta_{i_{m-1}}$ of the left-hand side of a decision rule. We denote this prefix by $\beta$, and by $T'$ we denote the table $T(f_{i_1}, \delta_{i_1}) \ldots (f_{i_{m-1}}, \delta_{i_{m-1}})$. If $P(T') \leq \alpha P(T)$ then we construct the $\alpha$-decision rule

$$\beta \to b$$

where $b$ is the most common decision for $T'$. We assign the decision $b$ to $\bar{\delta}$.

Let $P(T') > \alpha P(T)$ and $i_m$ be the minimum number from $\{1, \ldots, n\}$ for which $T'(f_{i_m}, \delta_{i_m})$ is nonempty and $P(T'(f_{i_m}, \delta_{i_m})) = \min P(T'(f_j, \delta_j))$ where minimum is considered among all $j \in \{1, \ldots, n\}$ for which $T'(f_j, \delta_j)$ is a nonempty table. If there are no such $j$ then we construct the decision rule

$$\beta \to b$$

where $b$ is the most common decision for $T'$. We assign the decision $b$ to $\bar{\delta}$. Otherwise, we form new prefix $\beta \wedge f_{i_m} = \delta_{i_m}$ and continue the work of algorithm.

We will say about this approach to prediction as about *lazy decision rules*. It should be noted that similar approaches are considered also as lazy decision trees [20].

*Example 7.7.* Let $T$ be the decision table depicted in Fig. 3.1. We apply "lazy decision rules" to $\alpha = 1/8$, $T$ and three tuples of values of attributes $f_1$, $f_2$, $f_3$ which are not rows of $T$ to predict values of the decision attribute $d$ (see Fig. 7.4).

$T =$

| $f_1$ | $f_2$ | $f_3$ | $d$ |
|---|---|---|---|
| 1 | 1 | 1 | 1 |
| 0 | 1 | 0 | 2 |
| 1 | 1 | 0 | 2 |
| 0 | 0 | 1 | 3 |
| 1 | 0 | 0 | 3 |

| | | | | |
|---|---|---|---|---|
| 0 | 0 | 0 | 3 | $f_2 = 0 \to 3$ |
| 0 | 1 | 1 | 2 | $f_1 = 0 \to 2$ |
| 1 | 0 | 1 | 3 | $f_2 = 0 \to 3$ |

**Fig. 7.4**

We have $P(T) = 8$, $P(T(f_1, 0)) = 1$, $P(T(f_1, 1)) = 3$, $P(T(f_2, 0)) = 0$, $P(T(f_2, 1)) = 2$, $P(T(f_3, 0)) = 2$, and $P(T(f_3, 1)) = 1$. Using this information it is easy to check that "lazy decision rules" algorithm constructs for the considered three tuples 1/8-decision rules depicted in Fig. 7.4.

### *7.3.3 Lazy Learning Algorithm Based on Decision Rules*

In this section, we consider the same classification (prediction) problem as in the previous sections: for a given decision table $T$ with $n$ columns labeled with attributes $f_1, \ldots, f_n$ and a new object $\bar{\delta}$ given by the tuple $(\delta_1, \ldots, \delta_n)$ of values of conditional attributes $f_1, \ldots, f_n$ we should generate the value of the decision attribute for $\bar{\delta}$.

We can construct a complete system of decision rules for $T$ and use it for prediction, or use "lazy decision rules" approach and construct a decision rule for $\bar{\delta}$ directly based on the decision table $T$.

In this section, we consider another way proposed and studied by J. Bazan [3, 4, 5]. For new object $\bar{\delta}$ and each decision $b$ from $T$, we find (using polynomial-time algorithm) the number $D(T, b, \bar{\delta})$ of rows $r$ from $T$ such that there exists a decision rule over $T$ which is true for $T$, realizable for $r$ and $\bar{\delta}$, and has $b$ on the right-hand side. For $\bar{\delta}$, we choose the minimum decision $b$ for which the value of $D(T, b, \bar{\delta})$ is maximum.

For a row $r = (r_1, \ldots, r_n)$ of $T$, we denote by $M(r, \bar{\delta})$ the set of attributes $f_i \in \{f_1, \ldots, f_n\}$ for which $r_i = \delta_i$.

**Proposition 7.8.** *A rule over $T$, which is true for $T$, realizable for $r$ and $\bar{\delta}$, and has $b$ on the right-hand side, exists if and only if the rule*

$$\bigwedge_{f_j \in M(r, \bar{\delta})} f_j = \delta_j \rightarrow b \tag{7.5}$$

*is true for $T$. If (7.5) is true for $T$ then (7.5) is a desired rule.*

*Proof.* Let (7.5) be true for $T$. It is clear that (7.5) is realizable for $r$ and $\bar{\delta}$, and has $b$ on the right-hand side. So (7.5) is a desired rule.

Let there exist a decision rule

$$f_{j_1} = b_1 \wedge \ldots \wedge f_{j_t} = b_t \rightarrow b \tag{7.6}$$

over $T$ which is true for $T$, realizable for $r$ and $\bar{\delta}$, and has $b$ on the right-hand side. Since this rule is realizable for $r$ and $\bar{\delta}$, we have $f_{j_1}, \ldots, f_{j_t} \in M(r, \bar{\delta})$. Since (7.6) is true for $T$, the rule (7.5) is true for $T$. □

Note that (7.5) is not true if $b$ is not the decision attached to the row $r$.

Now we have simple way for implementation of the classification algorithm described at the beginning of this section: $D(T, b, \bar{\delta})$ is equal to the number of rows $r$ from $T$ such that (i) the row $r$ is labeled with the decision $b$, and (ii) the rule (7.5) is true for $T$.

*Example 7.9.* Let us consider the decision table $T$ depicted in Fig. 3.1. We apply the lazy learning algorithm based on decision rules to the table $T$ and to three tuples of values of attributes $f_1$, $f_2$, $f_3$ which are not rows of $T$ to predict values of the decision attribute $d$ (see Fig. 7.5). Let us consider the work of our algorithm on the tuple $\bar{\delta} = (0, 0, 0)$. There are three possible decisions: 1, 2 and 3.

There is only one row in $T$ which is labeled with the decision 1. The rule (7.5) corresponding to this row and new object $\bar{\delta}$ is $\rightarrow 1$. This rule is not true for $T$. Therefore $D(T, 1, \bar{\delta}) = 0$.

$T =$

| $f_1$ | $f_2$ | $f_3$ | $d$ |
|---|---|---|---|
| 1 | 1 | 1 | 1 |
| 0 | 1 | 0 | 2 |
| 1 | 1 | 0 | 2 |
| 0 | 0 | 1 | 3 |
| 1 | 0 | 0 | 3 |

| | | | |
|---|---|---|---|
| 0 | 0 | 0 | 3 |
| 0 | 1 | 1 | 1 |
| 1 | 0 | 1 | 3 |

**Fig. 7.5**

There are two rows in $T$ which are labeled with the decision 2. The rules (7.5) corresponding to these rows and $\bar{\delta}$ are $f_1 = 0 \wedge f_3 = 0 \rightarrow 2$ and $f_3 = 0 \rightarrow 2$. The first rule is true for $T$, and the second one is not true for $T$. Therefore $D(T, 2, \bar{\delta}) = 1$.

There are two rows in $T$ which are labeled with the decision 3. The corresponding rules are $f_1 = 0 \wedge f_2 = 0 \rightarrow 3$ and $f_2 = 0 \wedge f_3 = 0 \rightarrow 3$. Both rules are true for $T$. Therefore $D(T, 3, \bar{\delta}) = 2$.

As a result, we attach to $\bar{\delta}$ the decision 3.

## *7.3.4 Lazy Learning Algorithm Based on Reducts*

It is very difficult to construct the whole set of true decision rules for a given decision table $T$ but we can efficiently extract some useful information about this set and based on this information predict value of the decision attribute for a new object given by values of conditional attributes (see Sect. 7.3.3 for details).

In this section, we consider similar approach but based on an information about the set of reducts. In general case, the number of reducts can grow exponentially with the growth of the number of columns and rows in decision tables. However, we can extract efficiently some information about the set of reducts and use this information for the prediction of decision attribute values.

Let $T$ be a decision table with $n$ columns labeled with attributes $f_1, \ldots, f_n$ that have values from the set $\{0, 1\}$. Let $\bar{\delta} = (\delta_1, \ldots, \delta_n)$ be a new object given by values of attributes $f_1, \ldots, f_n$.

We construct in polynomial time the canonical form $C(T)$ of the table $T$ (see Sect. 2.2.2) We know (see Lemma 2.7) that the set of rows of the table $C(T)$ with the exception of the first row coincides with the set of upper zeros $U_T$ of the characteristic function $f_T : \{0, 1\}^n \rightarrow \{0, 1\}$ corresponding to the table $T$.

Let $\bar{\beta} \in \{0, 1\}^n$ and $i_1, \ldots, i_m$ be numbers of digits from $\bar{\beta}$ which are equal to 1. Then $f_T(\bar{\beta}) = 1$ if and only if the set of attributes $\{f_{i_1}, \ldots, f_{i_m}\}$ is a test for $T$. We know (see Lemma 2.4) that the set of lower units of $f_T$ coincides with the set of tuples corresponding to reducts for the table $T$. It is too complicated for us to construct the set of lower units (reducts) but we can construct efficiently the set of upper zeros of $f_T$ which describes complectly the set of lower units.

The idea of lazy learning algorithm based on reducts is the following. For any decision $b$ from the decision table $T$, we add to $T$ the row $\bar{\delta}$ labeled with the decision $b$. As a result, we obtain new decision table $T(b)$. We construct the canonical form $C(T(b))$ of the table $T(b)$ and the set $U_{T(b)}$ of upper zeros of the characteristic function $f_{T(b)}$ corresponding to the table $T(b)$.

We find the number $|(U_T \cup U_{T(b)}) \setminus (U_T \cap U_{T(b)})|$ which will be considered as the distance between $U_T$ and $U_{T(b)}$. We assign to $\bar{\delta}$ a decision $b$ for which the distance between $U_T$ and $U_{T(b)}$ is minimum.

*Example 7.10.* Let us consider the decision table $T$ depicted in Fig. 7.6 and the new object $\bar{\delta} = (0, 0, 0)$. We construct tables $C(T)$, $T(1)$, $C(T(1))$, $T(2)$ and $C(T(2))$ (see Fig. 7.6).

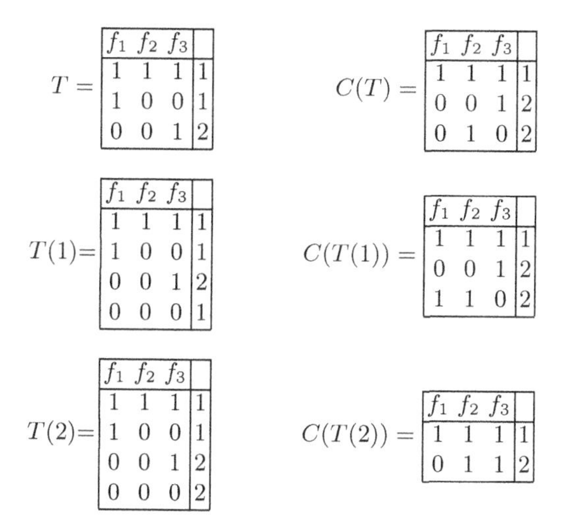

$T =$

| $f_1$ | $f_2$ | $f_3$ | |
|---|---|---|---|
| 1 | 1 | 1 | 1 |
| 1 | 0 | 0 | 1 |
| 0 | 0 | 1 | 2 |

$C(T) =$

| $f_1$ | $f_2$ | $f_3$ | |
|---|---|---|---|
| 1 | 1 | 1 | 1 |
| 0 | 0 | 1 | 2 |
| 0 | 1 | 0 | 2 |

$T(1) =$

| $f_1$ | $f_2$ | $f_3$ | |
|---|---|---|---|
| 1 | 1 | 1 | 1 |
| 1 | 0 | 0 | 1 |
| 0 | 0 | 1 | 2 |
| 0 | 0 | 0 | 1 |

$C(T(1)) =$

| $f_1$ | $f_2$ | $f_3$ | |
|---|---|---|---|
| 1 | 1 | 1 | 1 |
| 0 | 0 | 1 | 2 |
| 1 | 1 | 0 | 2 |

$T(2) =$

| $f_1$ | $f_2$ | $f_3$ | |
|---|---|---|---|
| 1 | 1 | 1 | 1 |
| 1 | 0 | 0 | 1 |
| 0 | 0 | 1 | 2 |
| 0 | 0 | 0 | 2 |

$C(T(2)) =$

| $f_1$ | $f_2$ | $f_3$ | |
|---|---|---|---|
| 1 | 1 | 1 | 1 |
| 0 | 1 | 1 | 2 |

**Fig. 7.6**

As a result, we have $U_T = \{(0, 0, 1), (0, 1, 0)\}$, $U_{T(1)} = \{(0, 0, 1), (1, 1, 0)\}$, and $U_{T(2)} = \{(0, 1, 1)\}$. The distance between $U_T$ and $U_{T(1)}$ is equal to 2. The distance between $U_T$ and $U_{T(2)}$ is equal to 3. Therefore we assign to $\bar{\delta}$ the decision 1.

## 7.4 Conclusions

The chapter is devoted to the consideration of the problem of supervised learning: for a given decision table $T$ and a tuple $\bar{\delta}$ of values of conditional attributes of $T$ which is not a row of $T$ we should predict the value of decision attribute for $\bar{\delta}$.

To this end, we either create a classifier (decision tree or decision rule system) which allows us to predict a decision for $\bar{\delta}$, or we can omit the construction of classifier and use the information contained in $T$ and $\bar{\delta}$ directly (lazy decision trees and rules, and lazy learning algorithms based on rules and reducts).

It is interesting to compare different results of prediction for different approaches (see Figs. 7.1–7.5). The variety of approaches increases the chance to find an appropriate way for prediction for a given decision table.

# 8

# Local and Global Approaches to Study of Trees and Rules

In this chapter, we consider two approaches to the study of decision trees and decision rule systems for problems over finite and infinite information systems. Local approach is based on the assumption that only attributes contained in a problem description are used in decision trees and decision rules systems solving this problem. Global approach is based on the assumption that any attributes from the considered information system can be used in decision trees and decision rule systems solving the problem.

The main difficulty in the global approach is the necessity to choose appropriate attributes in large or infinite set of attributes. However, in the frameworks of the global approach we can often construct more simple decision trees and decision rule systems rather than in the frameworks of the local approach.

This chapter is devoted to the study of growth in the worst case of time (and, sometimes, space) complexity of decision trees and decision rule systems with the growth of the number of attributes in problem description.

The chapter contains four sections. In Sect. 8.1, basic notions are discussed. Section 8.2 is devoted to the consideration of local approach for infinite and finite information systems. In Sect. 8.3, results related to the global approach are considered. Section 8.4 contains conclusions.

## 8.1 Basic Notions

Let $A$ be a nonempty set, $B$ be a finite nonempty set with at least two elements, and $F$ be a nonempty set of functions from $A$ to $B$. Functions from $F$ will be called *attributes* and the triple $U = (A, B, F)$ will be called an *information system*. If $F$ is a finite set then $U$ will be called a *finite* information system. If $F$ is an infinite set then $U$ will be called an *infinite* information system.

We will consider problems over the information system $U$. A *problem over* $U$ is an arbitrary $(n + 1)$-tuple $z = (\nu, f_1, \ldots, f_n)$ where $\nu : B^n \to \omega$,

M. Moshkov and B. Zielosko: Combinatorial Machine Learning, SCI 360, pp. 127–142.
springerlink.com © Springer-Verlag Berlin Heidelberg 2011

$\omega = \{0, 1, 2, \ldots\}$, and $f_1, \ldots, f_n \in F$. The number $\dim z = n$ will be called the *dimension* of the problem $z$. The problem $z$ may be interpreted as a problem of searching for the value $z(a) = \nu(f_1(a), \ldots, f_n(a))$ for an arbitrary $a \in A$. Different problems of pattern recognition, discrete optimization, fault diagnosis and computational geometry can be represented in such form. We denote by $\mathcal{P}(U)$ the set of all problems over the information system $U$.

As algorithms for problem solving we will consider decision trees and decision rule systems.

A *decision tree over* $U$ is a marked finite tree with the root in which each terminal node is labeled with a number from $\omega$; each node which is not terminal (such nodes are called *working*) is labeled with an attribute from $F$; each edge is labeled with an element from $B$. Edges starting in a working node are labeled with pairwise different elements.

Let $\Gamma$ be a decision tree over $U$. A *complete path* in $\Gamma$ is an arbitrary sequence $\xi = v_1, d_1, \ldots, v_m, d_m, v_{m+1}$ of nodes and edges of $\Gamma$ such that $v_1$ is the root, $v_{m+1}$ is a terminal node, and $v_i$ is the initial and $v_{i+1}$ is the terminal node of the edge $d_i$ for $i = 1, \ldots, m$. Now we define a system of equations $\mathcal{S}(\xi)$ and a subset $\mathcal{A}(\xi)$ of the set $A$ associated with $\xi$. If $m = 0$ then $\mathcal{S}(\xi)$ is empty system and $\mathcal{A}(\xi) = A$. Let $m > 0$, the node $v_i$ be labeled with the attribute $f_i$, and the edge $d_i$ be labeled with the element $\delta_i$ from $B$, $i = 1, \ldots, m$. Then $\mathcal{S}(\xi) = \{f_1(x) = \delta_1, \ldots, f_m(x) = \delta_m\}$ and $\mathcal{A}(\xi)$ is the set of solutions of $\mathcal{S}(\xi)$ from $A$.

We will say that a decision tree $\Gamma$ over $U$ *solves* a problem $z$ over $U$ if for any $a \in A$ there exists a complete path $\xi$ in $\Gamma$ such that $a \in \mathcal{A}(\xi)$, and the terminal node of the path $\xi$ is labeled with the number $z(a)$.

For decision trees, as time complexity measure we will consider the *depth* of a decision tree which is the maximum number of working nodes in a complete path in the tree. As space complexity measure we will consider the number of nodes in a decision tree. We denote by $h(\Gamma)$ the depth of a decision tree $\Gamma$, and by $\#(\Gamma)$ we denote the number of nodes in $\Gamma$. Note that for each problem $z$ over $U$ there exists a decision tree $\Gamma$ over $U$ which solves the problem $z$ and for which $h(\Gamma) \leq \dim z$ and $\#(\Gamma) \leq |B|^{\dim z + 1}$.

A *decision rule over* $U$ is an arbitrary expression of the kind

$$f_1 = \delta_1 \wedge \ldots \wedge f_m = \delta_m \rightarrow \sigma$$

where $f_1, \ldots, f_m \in F$, $\delta_1, \ldots, \delta_m \in B$ and $\sigma \in \omega$. Denote this decision rule by $\varrho$. The number $m$ will be called the *length* of the rule $\varrho$. Now we define a system of equations $\mathcal{S}(\varrho)$ and a subset $\mathcal{A}(\varrho)$ of the set $A$ associated with $\varrho$. If $m = 0$ then $\mathcal{S}(\varrho)$ is empty system and $\mathcal{A}(\varrho) = A$. Let $m > 0$. Then $\mathcal{S}(\varrho) = \{f_1(x) = \delta_1, \ldots, f_m(x) = \delta_m\}$ and $\mathcal{A}(\varrho)$ is the set of solutions of $\mathcal{S}(\varrho)$ from $A$. The number $\sigma$ will be called the *right-hand side of the rule* $\varrho$.

A *decision rule system over* $U$ is a nonempty finite set of decision rules over $U$. Let $S$ be a decision rule system over $U$ and $z$ be a problem over $U$. We will say that the decision rule system $S$ is *complete for the problem* $z$ if for any $a \in A$ there exists a rule $\varrho \in S$ such that $a \in \mathcal{A}(\varrho)$, and for each rule $\varrho \in S$ such that $a$ is a solution of $\mathcal{S}(\varrho)$, the right-hand side of $\varrho$ coincides with the number $z(a)$.

For decision rule systems, as time complexity measure we will consider the maximum length $L(S)$ of a rule from the system $S$. We will say about $L(S)$ as about the *depth* of decision rule system $S$. As space complexity measure we will consider the number of rules in a system. Note that for each problem $z$ over $U$ there exists a decision rule system $S$ over $U$ which is complete for the problem $z$ and for which $L(S) \leq \dim z$ and $|S| \leq |B|^{\dim z}$.

The investigation of decision trees and decision rule systems for a problem $z = (\nu, f_1, \ldots, f_n)$ which use only attributes from the set $\{f_1, \ldots, f_n\}$ is based on the study of the *decision table* $T(z)$ associated with the problem $z$. The table $T(z)$ is a rectangular table with $n$ columns which contains elements from $B$. The row $(\delta_1, \ldots, \delta_n)$ is contained in the table $T(z)$ if and only if the equation system

$$\{f_1(x) = \delta_1, \ldots, f_n(x) = \delta_n\}$$

is compatible on $A$ (has a solution from the set $A$). This row is labeled with the number $\nu(\delta_1, \ldots, \delta_n)$. For $i = 1, \ldots, n$, the $i$-th column is labeled with the attribute $f_i$. We know that a decision tree over $T(z)$ solves the problem $z$ if and only if this tree is a decision tree for $T(z)$. We know also that a decision rule system over $T(z)$ is complete for the problem $z$ if and only if this system is a complete decision rule system for $T(z)$.

If we would like to consider additional attributes $f_{n+1}, \ldots, f_{n+m}$, we can study a new problem $z' = (\mu, f_1, \ldots, f_n, f_{n+1}, \ldots, f_{n+m})$ such that

$$\mu(x_1, \ldots, x_{n+m}) = \nu(x_1, \ldots, x_n) ,$$

and corresponding decision table $T(z')$.

## 8.2 Local Approach to Study of Decision Trees and Rules

Let $U = (A, B, F)$ be an information system. For a problem $z = (\nu, f_1, \ldots, f_n)$ over $U$ we denote by $h_U^l(z)$ the minimum depth of a decision tree over $U$ which solves the problem $z$ and uses only attributes from the set $\{f_1, \ldots, f_n\}$. We denote by $L_U^l(z)$ the minimum depth of a complete decision rule system for $z$ which uses only attributes from the set $\{f_1, \ldots, f_n\}$. We will consider relationships among the parameters $h_U^l(z)$, $L_U^l(z)$ and $\dim z$. One can interpret the value $\dim z$ for the problem $z = (\nu, f_1, \ldots, f_n)$ as the depth of decision tree which solves the problem $z$ in trivial way by computing sequentially the values of the attributes $f_1, \ldots, f_n$. One can interpret also the value $\dim z$ for the problem $z = (\nu, f_1, \ldots, f_n)$ as the depth of trivial complete decision rule system for $z$ in which each rule is of the kind $f_1 = \delta_1 \wedge \ldots \wedge f_n = \delta_n \rightarrow \sigma$. So we will consider relationships between depth of locally optimal and trivial decision trees and decision rule systems. To this end, we define the functions $h_U^l : \omega \setminus \{0\} \to \omega$ and $L_U^l : \omega \setminus \{0\} \to \omega$ in the following way:

$$h_U^l(n) = \max\{h_U^l(z) : z \in \mathcal{P}(U), \dim z \le n\} ,$$
$$L_U^l(n) = \max\{L_U^l(z) : z \in \mathcal{P}(U), \dim z \le n\}$$

for any $n \in \omega \setminus \{0\}$, where $\mathcal{P}(U)$ is the set of all problems over $U$. The value $h_U^l(n)$ is the unimprovable upper bound on the value $h_U^l(z)$ for problems $z \in \mathcal{P}(U)$ such that $\dim z \le n$. The value $L_U^l(n)$ is the unimprovable upper bound on the value $L_U^l(z)$ for problems $z \in \mathcal{P}(U)$ such that $\dim z \le n$. The functions $h_U^l$ and $L_U^l$ will be called the *local Shannon functions* for the information system $U$. Using Corollary 2.28 we obtain $L_U^l(n) \le h_U^l(n)$ for any $n \in \omega \setminus \{0\}$.

### *8.2.1 Local Shannon Functions for Arbitrary Information Systems*

In [41] it was shown (see also [47, 53]) that for an arbitrary information system $U$ either $h_U^l(n) = O(1)$, or $h_U^l(n) = \Theta(\log_2 n)$, or $h_U^l(n) = n$ for each $n \in \omega \setminus \{0\}$.

The first type of behavior ($h_U^l(n) = O(1)$) realizes only for finite information systems.

The second type of behavior ($h_U^l(n) = \Theta(\log_2 n)$) is the most interesting for us since there exist two natural numbers $c_1$ and $c_2$ such that for each problem $z$ over $U$ the inequality $h_U^l(z) \le c_1 \log_2(\dim z) + c_2$ holds.

The third type of behavior ($h_U^l(n) = n$ for each $n \in \omega \setminus \{0\}$) is bad for us: for each $n \in \omega \setminus \{0\}$, there exists a problem $z$ over $U$ such that $h_U^l(z) = \dim z = n$. So, in the worst case the depth of locally optimal decision tree is equal to the depth of trivial decision tree.

Thus we must have possibility to discern the types of behavior. Now we consider the criterions of the local Shannon function $h_U^l$ behavior.

We will say that the information system $U = (A, B, F)$ satisfies the *condition of reduction* if there exists a number $m \in \omega \setminus \{0\}$ such that for each compatible on $A$ system of equations

$$\{f_1(x) = \delta_1, \dots, f_r(x) = \delta_r\}$$

where $r \in \omega \setminus \{0\}$, $f_1, \dots, f_r \in F$ and $\delta_1, \dots, \delta_r \in B$ there exists a subsystem of this system which has the same set of solutions and contains at most $m$ equations.

In the following theorem, the criterions of the local Shannon function $h_U^l$ behavior are considered.

**Theorem 8.1.** ([41]) *Let $U$ be an information system. Then the following statements hold:*

*a) if $U$ is a finite information system then $h_U^l(n) = O(1)$;*

*b) if $U$ is an infinite information system which satisfies the condition of reduction then $h_U^l(n) = \Theta(\log_2 n)$;*

*c) if $U$ is an infinite information system which does not satisfy the condition of reduction then $h_U^l(n) = n$ for each $n \in \omega \setminus \{0\}$.*

There are only two types of behavior of the local Shannon function $L_U^l$.

**Theorem 8.2.** *Let $U = (A, B, F)$ be an information system. Then the following statements hold:*

*a) if $U$ is a finite information system then $L_U^l(n) = O(1)$;*

*b) if $U$ is an infinite information system which satisfies the condition of reduction then $L_U^l(n) = O(1)$;*

*c) if $U$ is an infinite information system which does not satisfy the condition of reduction then $L_U^l(n) = n$ for each $n \in \omega \setminus \{0\}$.*

*Proof.* We know that $L_U^l(n) \leq h_U^l(n)$ for any $n \in \omega \setminus \{0\}$. By Theorem 8.1, if $U$ is a finite information system then $L_U^l(n) = O(1)$.

Let $U$ be an infinite information system which satisfies the condition of reduction. Then there exists a number $m \in \omega \setminus \{0\}$ such that for each compatible on $A$ system of equations

$$\{f_1(x) = \delta_1, \ldots, f_r(x) = \delta_r\} ,$$

where $r \in \omega \setminus \{0\}$, $f_1, \ldots, f_r \in F$ and $\delta_1, \ldots, \delta_r \in B$, there exists a subsystem of this system which has the same set of solutions and contains at most $m$ equations. From here it follows that for any problem $z$ over $U$ and for any row $\bar{\delta}$ of the decision table $T(z)$ the inequality $M(T(z), \bar{\delta}) \leq m$ holds. From Theorem 3.11 it follows that $L(T(z)) \leq m$. Therefore there exists a decision rule system $S$ over $U$ which is complete for the problem $z$, use only attributes from the description of $z$ and for which $L(S) \leq m$. Thus, $L_U^l(n) \leq m$ for any $n \in \omega \setminus \{0\}$, and $L_U^l(n) = O(1)$.

Let $U$ be an infinite information system which does not satisfy the condition of reduction. By Theorem 8.1, $h_U^l(n) = n$ and therefore $L_U^l(n) \leq n$ for any $n \in \omega \setminus \{0\}$.

It is not difficult to show that for any $n \in \omega \setminus \{0\}$, there exists a compatible on $A$ system of equations

$$\{f_1(x) = \delta_1, \ldots, f_n(x) = \delta_n\} ,$$

where $f_1, \ldots, f_n \in F$ and $\delta_1, \ldots, \delta_n \in B$, such that for any proper subsystem of this system the set of solutions of the subsystem on $A$ is different from the set of solutions of the initial system on $A$.

We consider the problem $z = (\nu, f_1, \ldots, f_n)$ such that $\nu(\bar{\delta}) = 1$ where $\bar{\delta} = (\delta_1, \ldots, \delta_n)$ and for any row $\bar{\sigma}$ of the table $T(z)$ different from $\bar{\delta}$, $\nu(\bar{\sigma}) =$

2. One can show that $M(T(z), \bar{\delta}) = n$. From Theorem 3.11 it follows that $L(T(z)) = n$. Therefore $L^l_U(n) \geq n$ and $L^l_U(n) = n$. □

Now we consider an example.

*Example 8.3.* Let $m, t \in \omega \setminus \{0\}$. We denote by $Pol(m)$ the set of all polynomials which have integer coefficients and depend on variables $x_1, \ldots, x_m$. We denote by $Pol(m, t)$ the set of all polynomials from $Pol(m)$ such that the degree of each polynomial is at most $t$. We define information systems $U(m)$ and $U(m, t)$ as follows: $U(m) = (\mathbb{R}^m, E, F(m))$ and $U(m, t) = (\mathbb{R}^m, E, F(m, t))$ where $E = \{-1, 0, +1\}$, $F(m) = \{\text{sign}(p) : p \in Pol(m)\}$ and $F(m, t) = \{\text{sign}(p) : p \in Pol(m, t)\}$. Here $\text{sign}(x) = -1$ if $x < 0$, $\text{sign}(x) = 0$ if $x = 0$, and $\text{sign}(x) = +1$ if $x > 0$. One can prove that $h^l_{U(m)}(n) = L^l_{U(m)}(n) = n$ for each $n \in \omega \setminus \{0\}$, $h^l_{U(1,1)}(n) = \Theta(\log_2 n)$, $L^l_{U(1,1)}(n) = O(1)$, and if $m > 1$ or $t > 1$ then $h^l_{U(m,t)}(n) = L^l_{U(m,t)}(n) = n$ for each $n \in \omega \setminus \{0\}$.

### 8.2.2 Restricted Binary Information Systems

An information system $U = (A, B, F)$ is called *binary* if $B = \{0, 1\}$. A binary information system $U = (A, \{0, 1\}, F)$ is called *restricted* it it satisfies the condition of reduction.

Let $U = (A, \{0, 1\}, F)$ be a restricted binary information system. Then there exists a number $m \in \omega \setminus \{0\}$ such that for each compatible on $A$ system of equations

$$\{f_1(x) = \delta_1, \ldots, f_r(x) = \delta_r\} ,$$

where $r \in \omega \setminus \{0\}$, $f_1, \ldots, f_r \in F$ and $\delta_1, \ldots, \delta_r \in \{0, 1\}$, there exists a subsystem of this system which has the same set of solutions and contains at most $m$ equations.

Let $z = (\nu, f_1, \ldots, f_n)$ be a problem over $U$ and $T(z)$ be the decision table corresponding to this problem.

Let us show that

$$M(T(z)) \leq m + 1 . \tag{8.1}$$

Let $\bar{\delta} = (\delta_1, \ldots, \delta_n) \in \{0, 1\}^n$ and $\bar{\delta}$ be a row of $T(z)$. Then the system of equations

$$\{f_1(x) = \delta_1, \ldots, f_n(x) = \delta_n\} \tag{8.2}$$

is compatible on $A$ and there exists a subsystem of this system which has the same set of solutions and contains at most $m$ equations. From here it follows that $M(T(z), \bar{\delta}) \leq m$. Let $\bar{\delta}$ be not a row of $T(z)$. Then the system (8.2) is not compatible on $A$. Let us show that this system has a subsystem with at most $m + 1$ equations which is not compatible on $A$. If the system $\{f_1(x) = \delta_1\}$ is not compatible on $A$ then the considered statement holds. Let this system is compatible on $A$. Then there exists a number $t$ from $\{1, \ldots, n - 1\}$ such that the system

$$H = \{f_1(x) = \delta_1, \ldots, f_t(x) = \delta_t\}$$

is compatible on $A$ but the system $H \cup \{f_{t+1}(x) = \delta_{t+1}\}$ is incompatible on $A$. Then there exists a subsystem $H'$ of the system $H$ which has the same set of solutions and contains at most $m$ equations. Then $H' \cup \{f_{t+1}(x) = \delta_{t+1}\}$ is a subsystem of the system (8.2) which has at most $m+1$ equations and is not compatible on $A$. Therefore $M(T(z), \bar{\delta}) \leq m+1$ and the inequality (8.1) holds.

Let $\alpha$ be a real number such that $0 < \alpha < 1$. From (8.1) and Theorem 6.19 it follows that the minimum depth of $\alpha$-decision tree for the table $T(z)$ is at most $(m+1)(\log_2(1/\alpha)+1)$. From (8.1) and Theorem 3.11 it follows that $L(T(z)) \leq m+1$. From Theorem 6.30 it follows that for $\alpha$-complete decision rule system $S$ for $T(z)$ constructed by the greedy algorithm, the inequality $L(S) \leq (m+1)\ln(1/\alpha)+1$ holds. From Theorem 6.37 it follows that the depth of $\alpha$-decision tree for the table $T(z)$ constructed by the greedy algorithm is at most $(m+1)\ln(1/\alpha)+1$. Note that the obtained bounds do not depend on the dimension of problems.

Let us show that

$$N(T(z)) \leq 2^m n^m \,. \tag{8.3}$$

We know that for any system of equations of the kind

$$\{f_1(x) = \delta_1, \ldots, f_n(x) = \delta_n\} \,,$$

where $\delta_1, \ldots, \delta_n \in \{0,1\}$, which is compatible on $A$ there exists a subsystem with at most $m$ equations that has the same set of solutions. One can show that the number of systems of equations of the kind

$$\{f_{i_1}(x) = \delta_1, \ldots, f_{i_t}(x) = \delta_t\} \,,$$

where $1 \leq t \leq m$, $f_{i_1}, \ldots, f_{i_t} \in \{f_1, \ldots, f_n\}$ and $\delta_1, \ldots, \delta_t \in \{0,1\}$ is at most $2^m n^m$. Therefore the equation (8.3) holds. From (8.1), (8.3) and Theorem 3.17 it follows that the minimum depth of a decision tree for the table $T(z)$ is at most $m(m+1)(\log_2 n+1)$. So, we have at most logarithmic growth of the minimum depth of decision trees depending on the dimension of problems.

Let us show that

$$|SEP(T(z))| \leq 2^m n^m + 1 \,, \tag{8.4}$$

where $SEP(T(z))$ is the set of all separable subtables of the table $T(z)$ including $T(z)$. Each separable subtable of $T(z)$ is a nonempty subtable of $T(z)$ which can be represented in the form $T = T(z)(f_{i_1}, \delta_{i_1}) \ldots (f_{i_r}, \delta_{i_r})$ where $r \in \{0, 1, \ldots, n\}$, $f_{i_1}, \ldots, f_{i_r} \in \{f_1, \ldots, f_n\}$ and $\delta_{i_1}, \ldots, \delta_{i_r} \in \{0,1\}$. Let $r > 0$. Since the considered subtable is nonempty, the system of equations

$$\{f_{i_1}(x) = \delta_{i_1}, \ldots, f_{i_r}(x) = \delta_{i_r}\}$$

is compatible on $A$, and there exists a subsystem

$$\{f_{j_1}(x) = \delta_{j_1}, \ldots, f_{j_t}(x) = \delta_{j_t}\}$$

of this system, which has the same set of solutions and for which $t \leq m$. It is clear that $T = T(z)(f_{j_1}, \delta_{j_1}) \ldots (f_{j_t}, \delta_{j_t})$. Using this fact it is not difficult to show that the number of separable subtables of the table $T(z)$ which are not equal to $T(z)$ is at most $2^m n^m$. As a result, we have $|SEP(T(z))| \leq 2^m n^m + 1$.

From (8.3) and (8.4) and from Theorems 4.23, 4.26, 6.42, and 6.43 it follows that algorithms $W$, $V$, $W_\alpha$, and $V_\alpha$ for exact optimization of decision and $\alpha$-decision trees and rules have on tables $T(z)$ polynomial time complexity depending on $\dim z$.

The mentioned facts determine the interest of studying restricted information systems.

We now consider restricted binary linear information systems. Let $P$ be the set of all points in the plane and $l$ be a straight line (line in short) in the plane. This line divides the plane into two open half-planes $H_1$ and $H_2$ and the line $l$. Two attributes *correspond* to the line $l$. The first attribute takes value 0 on points from $H_1$, and value 1 on points from $H_2$ and $l$. The second one takes value 0 on points from $H_2$, and value 1 on points from $H_1$ and $l$. Denote by $\mathcal{L}$ the set of all attributes corresponding to lines in the plane. Information systems of the kind $(P, \{0, 1\}, F)$, where $F \subseteq \mathcal{L}$, will be called *binary linear information systems in the plane.* We will describe all restricted binary linear information systems in the plane.

Let $l$ be a line in the plane. Denote by $\mathcal{L}(l)$ the set of all attributes corresponding to lines which are parallel to $l$ (see Fig. 8.1). Let $p$ be a point in the plane. Denote by $\mathcal{L}(p)$ the set of all attributes corresponding to lines which pass through $p$ (see Fig. 8.2). A set $C$ of attributes from $\mathcal{L}$ will be called a *clone* if $C \subseteq \mathcal{L}(l)$ for some line $l$ or $C \subseteq \mathcal{L}(p)$ for some point $p$.

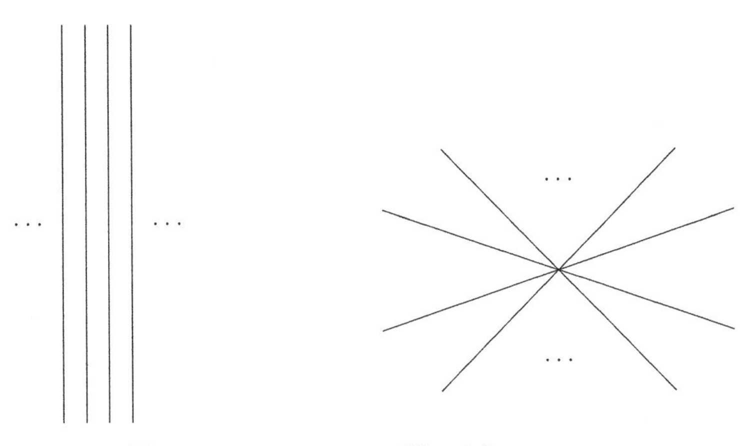

**Fig. 8.1** **Fig. 8.2**

Proof of the next statement can be found in [54].

**Theorem 8.4.** *A binary linear information system in the plane $(P, \{0,1\}, F)$ is restricted if and only if $F$ is the union of a finite number of clones.*

### *8.2.3 Local Shannon Functions for Finite Information Systems*

Theorems 8.1 and 8.2 give us some information about the behavior of local Shannon functions for infinite information systems. But for a finite information system $U$ we have only the relations $h^l_U(n) = O(1)$ and $L^l_U(n) = O(1)$. However, finite information systems are important for different applications.

Now we consider the behavior of the local Shannon functions for an arbitrary finite information system $U = (A, B, F)$ such that $f \not\equiv \text{const}$ for any $f \in F$.

A set $\{f_1, \ldots, f_n\} \subseteq F$ will be called *redundant* if $n \geq 2$ and there exist $i \in \{1, \ldots, n\}$ and $\mu : B^{n-1} \to B$ such that

$$f_i(a) = \mu(f_1(a), \ldots, f_{i-1}(a), f_{i+1}(a), \ldots, f_n(a))$$

for each $a \in A$. If the set $\{f_1, \ldots, f_n\}$ is not redundant then it will be called *irredundant*. We denote by $\text{ir}(U)$ the maximum number of attributes in an irredundant subset of the set $F$.

A *systems of equations over* $U$ is an arbitrary system

$$\{f_1(x) = \delta_1, \ldots, f_n(x) = \delta_n\} \tag{8.5}$$

such that $n \in \omega \setminus \{0\}$, $f_1, \ldots, f_n \in F$ and $\delta_1, \ldots, \delta_n \in B$. The system (8.5) will be called *cancelable* if $n \geq 2$ and there exists a number $i \in \{1, \ldots, n\}$ such that the system

$$\{f_1(x) = \delta_1, \ldots, f_{i-1}(x) = \delta_{i-1}, f_{i+1}(x) = \delta_{i+1}, \ldots, f_n(x) = \delta_n\}$$

has the same set of solutions just as the system (8.5). If the system (8.5) is not cancelable then it will be called *uncancelable*. We denote by $\text{un}(U)$ the maximum number of equations in an uncancelable compatible system over $U$.

One can show that

$$1 \leq \text{un}(U) \leq \text{ir}(U) .$$

The values $\text{un}(U)$ and $\text{ir}(U)$ will be called *the first and the second local critical points of the information system* $U = (A, B, F)$. Now we describe the behaviors of the local Shannon functions in terms of local critical points of $U$ and the cardinality of the set $B$. The next two theorems follow immediately from results obtained in [43] for problems with many-valued decisions.

**Theorem 8.5.** *Let $U = (A, B, F)$ be a finite information system such that $f \not\equiv \text{const}$ for any $f \in F$, and $n \in \omega \setminus \{0\}$. Then the following statements hold:*

*a) if $n \leq \text{un}(U)$ then $h^l_U(n) = n$;*

*b) if $\text{un}(U) \leq n \leq \text{ir}(U)$ then*

$$\max\{\text{un}(U), \log_k(n+1)\} \leq h^l_U(n) \leq \min\{n, 2(\text{un}(U))^2 \log_2 2(kn+1)\}$$

*where $k = |B|$;*

*c) if $n \geq \text{ir}(U)$ then $h^l_U(n) = h^l_U(\text{ir}(U))$.*

**Theorem 8.6.** *Let $U = (A, B, F)$ be a finite information system such that $f \not\equiv \text{const}$ for any $f \in F$, and $n \in \omega \setminus \{0\}$. Then the following statements hold:*

*a) if $n \leq \text{un}(U)$ then $L^l_U(n) = n$;*

*b) if $n \geq \text{un}(U)$ then $L^l_U(n) = \text{un}(U)$.*

Of course, the problem of computing the values $\text{un}(U)$ and $\text{ir}(U)$ for a given finite information system $U$ is very complicated problem. But obtained results allow us to constrict essentially the class of possible types of local Shannon functions for finite information systems.

*Example 8.7.* Denote by $P$ the set of all points in the plane. Consider an arbitrary straight line $l$, which divides the plane into positive and negative open half-planes, and the line $l$ itself. Assign a function $f : P \rightarrow \{0, 1\}$ to the line $l$. The function $f$ takes the value 1 if a point is situated on the positive half-plane, and $f$ takes the value 0 if a point is situated on the negative half-plane or on the line $l$. Denote by $F$ the set of functions which correspond to certain $r$ mutually disjoint finite classes of parallel straight lines. Consider a finite information system $U = (P, \{0, 1\}, F)$. One can show that $\text{ir}(U) = |F|$ and $\text{un}(U) \leq 2r$.

## 8.3 Global Approach to Study of Decision Trees and Rules

First, we consider arbitrary infinite information systems. Later, we will study two-valued finite information systems.

### *8.3.1 Infinite Information Systems*

We will study not only time but also space complexity of decision trees and decision rule systems. It is possible to consider time and space complexity independently. For an arbitrary infinite information system, in the worst case, the following observations can be made (see [50]):

- the minimum depth of decision tree (as a function on the number of attributes in a problem description) either is bounded from below by logarithm and from above by logarithm to the power $1 + \varepsilon$, where $\varepsilon$ is an arbitrary positive constant, or grows linearly;
- the minimum depth of complete decision rule system (as a function on the number of attributes in a problem description) either is bounded from above by a constant or grows linearly;
- the minimum number of nodes in decision tree (as a function on the number of attributes in a problem description) has either polynomial or exponential growth;
- the minimum number of rules in complete decision rule system (as a function on the number of attributes in a problem description) has either polynomial or exponential growth.

The joint consideration of time and space complexity is more interesting for us. In this section, a partition of the set of all infinite information systems into two classes is considered. Information systems from the first class are close to the best ones from the point of view of time and space complexity of decision trees and decision rule systems. Decision trees and decision rule systems for information systems from the second class have in the worst case large time or space complexity.

For information systems from the first class, the following observations can be made (see [50]):
- there exist decision trees whose depth grows almost as logarithm, and the number of nodes grows almost as a polynomial on the number of attributes in a problem description;
- there exist complete decision rule systems whose depth is bounded from above by a constant, and the number of rules grows almost as a polynomial on the number of attributes in a problem description.

For an arbitrary information system from the second class in the worst case, the following observations can be made (see [50]):
- the minimum depth of decision tree (as a function on the number of attributes in a problem description) grows linearly;
- complete decision rule systems have at least linear growth of the depth or have at least exponential growth of the number of rules (depending on the number of attributes in a problem description).

### Partition of the Set of Infinite Information Systems

Let $U = (A, B, F)$ be an infinite information system.

Define the notion of *independence dimension* (or, in short, *I-dimension*) of information system $U$. A finite subset $\{f_1, \ldots, f_p\}$ of the set $F$ is called an *independent set* if there exist two-element subsets $B_1, \ldots, B_p$ of the set $B$ such that for any $\delta_1 \in B_1, \ldots, \delta_p \in B_p$ the system of equations

$$\{f_1(x) = \delta_1, \ldots, f_p(x) = \delta_p\} \tag{8.6}$$

is compatible on the set $A$ (has a solution from $A$). If for any natural $p$ there exists a subset of the set $F$, which cardinality is equal to $p$ and which is an independent set, then we will say that the information system $U$ has infinite I-dimension. Otherwise, I-dimension of $U$ is the maximum cardinality of a subset of $F$, which is an independent set.

The notion of I-dimension is closely connected with well known notion of Vapnik-Chervonenkis dimension [87]. In particular, an information system $(A, \{0, 1\}, F)$ has finite I-dimension if and only if it has finite VC-dimension [30].

Now we consider the condition of decomposition for the information system $U$. Let $p \in \omega \setminus \{0\}$. A nonempty subset $D$ of the set $A$ will be called $(p, U)$-set if $D$ coincides with the set of solutions on $A$ of a system of the kind (8.6) where $f_1, \ldots, f_p \in F$ and $\delta_1, \ldots, \delta_p \in B$ (we admit that among the attributes $f_1, \ldots, f_p$ there are identical ones).

We will say that the information system $U$ satisfies the *condition of decomposition* if there exist numbers $m, t \in \omega \setminus \{0\}$ such that every $(m + 1, U)$-set is a union of $t$ sets each of which is an $(m, U)$-set (we admit that among the considered $t$ sets there are identical ones).

We consider partition of the set of infinite information systems into two classes: $\mathcal{C}_1$ and $\mathcal{C}_2$. The class $\mathcal{C}_1$ consists of all infinite information systems each of which has finite I-dimension and satisfies the condition of decomposition. The class $\mathcal{C}_2$ consists of all infinite information systems each of which has infinite I-dimension or does not satisfy the condition of decomposition.

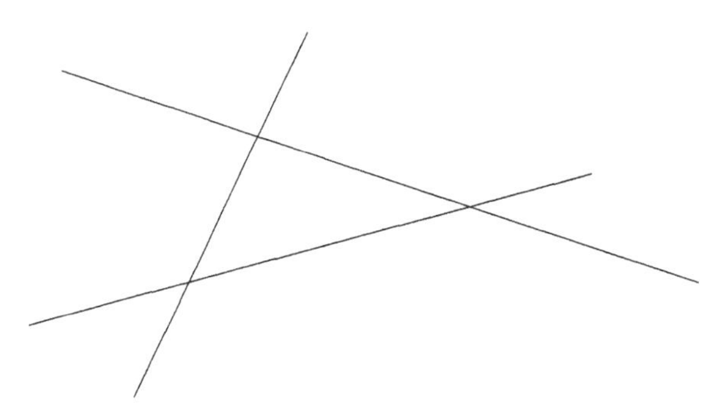

**Fig. 8.3**

We now consider an example of information system from the class $\mathcal{C}_1$.

*Example 8.8.* Let $P$ be the set of all points in the plane and $l$ be a straight line in the plane. This line divides the plane into two open half-planes $H_1$ and $H_2$ and the line $l$. We correspond one attribute to the line $l$. This attribute takes value 0 on points from $H_1$, and value 1 on points from $H_2$ and $l$. Denote by $\mathcal{F}$ the set of all attributes corresponding to lines in the plane. Let us consider the information system $\mathcal{U} = (P, \{0, 1\}, \mathcal{F})$.

The information system $\mathcal{U}$ has finite I-dimension: there are no three lines which divide the plane into eight domains (see Fig. 8.3). The information system $\mathcal{U}$ satisfies the condition of decomposition: each $(4,\mathcal{U})$-set is a union of two $(3,\mathcal{U})$-sets (see Fig. 8.4). Therefore $\mathcal{U} \in \mathcal{C}_1$.

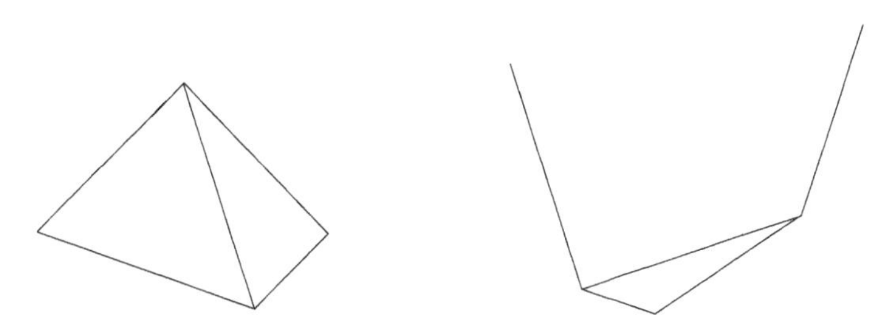

**Fig. 8.4**

## Bounds on Time and Space Complexity

In this section, we consider three theorems and an example from [50]. In the following theorem, time and space complexity of decision trees are considered.

**Theorem 8.9.** *Let $U = (A, B, F)$ be an infinite information system. Then the following statements hold:*

*a) if $U \in \mathcal{C}_1$ then for any $\varepsilon$, $0 < \varepsilon < 1$, there exists a positive constant $c$ such that for each problem $z$ over $U$ there exists a decision tree $\Gamma$ over $U$ which solves the problem $z$ and for which $h(\Gamma) \leq c(\log_2 n)^{1+\varepsilon} + 1$ and $\#(\Gamma) \leq |B|^{c(\log_2 n)^{1+\varepsilon}+2}$ where $n = \dim z$;*

*b) if $U \in \mathcal{C}_1$ then for any $n \in \omega \setminus \{0\}$ there exists a problem $z$ over $U$ with $\dim z = n$ such that for each decision tree $\Gamma$ over $U$, which solves the problem $z$, the inequality $h(\Gamma) \geq \log_{|B|}(n+1)$ holds;*

*c) if $U \in \mathcal{C}_2$ then for any $n \in \omega \setminus \{0\}$ there exists a problem $z$ over $U$ with $\dim z = n$ such that for each decision tree $\Gamma$ over $U$, which solves the problem $z$, the inequality $h(\Gamma) \geq n$ holds.*

In the next theorem, time and space complexity of decision rule systems are considered.

**Theorem 8.10.** *Let $U = (A, B, F)$ be an infinite information system. Then the following statements hold:*

*a) if $U \in \mathcal{C}_1$ then for any $\varepsilon$, $0 < \varepsilon < 1$, there exist positive constants $c_1$ and $c_2$ such that for each problem $z$ over $U$ there exists a decision rule system $S$ over $U$ which is complete for the problem $z$ and for which $L(S) \leq c_1$ and $|S| \leq |B|^{c_2(\log_2 n)^{1+\varepsilon}+1}$ where $n = \dim z$;*

*b) if $U \in \mathcal{C}_2$ then for any $n \in \omega \setminus \{0\}$ there exists a problem $z$ over $U$ with* $\dim z = n$ *such that for each decision rule system $S$ over $U$, which is complete for the problem $z$, the inequality $L(S) \geq n$ holds or the inequality $|S| \geq 2^n$ holds.*

In the following theorem, bounds are considered in which instead of $\varepsilon$ a function $\varphi(n)$ stands that decreases with the growth of $n$.

**Theorem 8.11.** *Let $U = (A, B, F)$ be an infinite information system from the class $\mathcal{C}_1$. Then there exist positive constants $c_1, c_2, c_3, c_4, c_5$ such that for the function $\varphi(n) = c_1/\sqrt{\log_2 \log_2 n}$ for any problem $z$ over $U$ with* $\dim z = n \geq c_2$ *the following statements hold:*

*a) there exists a decision tree $\Gamma$ over $U$ which solves $z$ and for which $h(\Gamma) \leq c_3(\log_2 n)^{1+\varphi(n)}$ and $\#(\Gamma) \leq |B|^{c_3(\log_2 n)^{1+\varphi(n)}+1}$;*

*b) there exists a decision rule system $S$ over $U$ which is complete for $z$ and for which $L(S) \leq c_4$ and $|S| \leq |B|^{c_5(\log_2 n)^{1+\varphi(n)}}$.*

So the class $\mathcal{C}_1$ is interesting from the point of view of different applications. The following example characterizes both the wealth and the boundedness of this class.

*Example 8.12.* Let $m, t \in \omega \setminus \{0\}$. We consider the same information systems $U(m)$ and $U(m, t)$ as in Example 8.3. One can prove that $U(m) \in \mathcal{C}_2$ and $U(m, t) \in \mathcal{C}_1$. Note that the system $U(m)$ has infinite I-dimension.

### 8.3.2 Global Shannon Function $h_U^l$ for Two-Valued Finite Information Systems

An information system $U = (A, B, F)$ will be called *two-valued* if $|B| = 2$. For a problem $z = (\nu, f_1, \ldots, f_n)$ over $U$ we denote by $h_U^g(z)$ the minimum depth of a decision tree over $U$ which solves the problem $z$. We will consider the relationships between the parameters $h_U^g(z)$ and $\dim z$. Recall that one can interpret the value $\dim z$ for the problem $z$ as the depth of the decision tree which solves the problem $z$ in trivial way by computing sequentially the values of attributes $f_1, \ldots, f_n$. We define the function $h_U^g : \omega \setminus \{0\} \to \omega$ in the following way:

$$h_U^g(n) = \max\{h_U^g(z) : z \in \mathcal{P}(U), \dim z \leq n\}$$

for any $n \in \omega \setminus \{0\}$. The value $h_U^g(n)$ is the unimprovable upper bound on the value $h_U^g(z)$ for problems $z \in \mathcal{P}(U)$ such that $\dim z \leq n$. The function $h_U^g$ will be called a *global Shannon function* for the information system $U$.

Now we consider the behavior of this global Shannon function for an arbitrary two-valued finite information system $U = (A, B, F)$ such that $f \not\equiv \text{const}$ for any $f \in F$.

Recall that by $\mathrm{ir}(U)$ we denote the maximum number of attributes in an irredundant subset of the set $F$ (see Sect. 8.2.3).

A problem $z \in P(U)$ will be called *stable* if $h_U^g(z) = \dim z$. We denote by $\mathrm{st}(U)$ the maximum dimension of a stable problem over $U$.

One can show that

$$1 \leq \mathrm{st}(U) \leq \mathrm{ir}(U) .$$

The values $\mathrm{st}(U)$ and $\mathrm{ir}(U)$ will be called *the first and the second global critical points of the information system* $U$. Now we describe the behavior of the global Shannon function $h_U^g$ in terms of global critical points of $U$.

**Theorem 8.13.** ([44]) *Let $U$ be a two-valued finite information system such that $f \not\equiv \mathrm{const}$ for any $f \in F$, and $n \in \omega \setminus \{0\}$. Then the following statements hold:*

*a) if $n \leq \mathrm{st}(U)$ then $h_U^g(n) = n$;*
*b) if $\mathrm{st}(U) < n \leq \mathrm{ir}(U)$ then*

$$\max\{\mathrm{st}(U), \log_2(n+1)\} \leq h_U^g(n) \leq \min\left\{n, 8(\mathrm{st}(U)+1)^5(\log_2 n)^2\right\} ;$$

*c) if $n \geq \mathrm{ir}(U)$ then $h_U^g(n) = h_U^g(\mathrm{ir}(U))$.*

The problem of computing the values $\mathrm{st}(U)$ and $\mathrm{ir}(U)$ for a given two-valued finite information system $U$ is a complicated problem. However, the obtained results allow us to constrict the class of possible types of the global Shannon function $h_U^g$.

*Example 8.14.* Let us consider the same information system $U = (P, \{0,1\}, F)$ as in Example 8.7. One can show that $\mathrm{st}(U) \leq 2r$ and $\mathrm{ir}(U) = |F|$.

## 8.4 Conclusions

This chapter is devoted to the consideration of local and global approaches to the study of decision trees and decision rule systems for problems over finite and infinite information systems. Our main aim is to describe the behavior in the worst case of minimum time and, sometimes, space complexity of decision trees and decision rule systems depending on the number of attributes in the problem description.

In the case of local approach, proofs of the considered results are based on lower and upper bounds on time complexity of decision trees and decision rule systems considered in the first part of this book. In the case of global approach, we should use also more advanced tools.

Note that an essential part of the considered results can be generalized to the case of problems with many-valued decisions $z = (\nu, f_1, \dots, f_n)$ where $\nu$ corresponds to each tuple of attribute $f_1, \dots, f_n$ values not a unique decision but a set of decisions. Our aim is to find at least one decision from this set.

One can show that the most complicated problems of a given dimension with one-valued decisions and with many-valued decisions have the same time complexity (minimum depth of a decision tree solving a problem and minimum depth of a complete decision rule system for a problem). Therefore Shannon functions in the case of one-valued decisions and in the case of many-valued decisions are the same.

In particular, the most complicated problem with one-valued decisions of the kind $z = (\nu, f_1, \ldots, f_n)$ is a problem for which the values of $\nu$ (there are numbers) for different inputs are different. The most complicated problem with many-valued decisions of the kind $z = (\nu, f_1, \ldots, f_n)$ is a problem for which values of $\nu$ (there are sets) for different inputs are disjoint.

# 9

# Decision Trees and Rules over Quasilinear Information Systems

In this chapter, we consider decision trees and decision rule systems over linear and quasilinear information systems, with applications to discrete optimization problems and analysis of acyclic programs in the basis $B_0 = \{x + y, x - y, 1; \text{sign}(x)\}$.

Each problem over a linear information system can be represented in the following form. We take finite number of hyperplanes in the space $\mathbb{R}^m$. These hyperplanes divide given part $C$ of the space into domains. These domains are numbered such that different domains can have the same number. For a given point from $C$, it is required to recognize the number of domain which contains this point. Decision trees and decision rules over the considered information system use attributes of the kind $\text{sign}\left(\sum_{i=1}^{m} a_i x_i + x_{m+1}\right)$. This attribute allows us to recognize the position of a point relative to the hyperplane defined by the equality $\sum_{i=1}^{m} a_i x_i + x_{m+1} = 0$.

Quasilinear information system is simple and useful generalization of linear information system: instead of attributes of the kind $\text{sign}\left(\sum_{i=1}^{m} a_i x_i + x_{m+1}\right)$ we consider attributes in the form $\text{sign}\left(\sum_{i=1}^{m} a_i \varphi_i + x_{m+1}\right)$ where $\varphi_1, \ldots, \varphi_m$ are functions from a set $A$ to $\mathbb{R}$.

Upper bounds on the depth of decision trees and complete systems of decision rules over quasilinear information systems are discussed in this chapter. We consider also a draft of the proof of similar bounds for a linear information system. This is an illustration of the use of tools from the first part of the book for the analysis of infinite information systems.

In the chapter, we consider two areas of applications of discussed results. We deal with three classes of optimization problems, and study relationships between depth of deterministic and nondeterministic acyclic programs in the basis $B_0 = \{x + y, x - y, 1; \text{sign}(x)\}$.

This chapter consists of four sections. Section 9.1 is devoted to the study of bounds on complexity of decision trees and rules over linear and quasilinear information systems. In Sect. 9.2, three classes of optimization problems over quasilinear information systems are discussed. In Sect. 9.3, acyclic programs

M. Moshkov and B. Zielosko: Combinatorial Machine Learning, SCI 360, pp. 143–153.
springerlink.com © Springer-Verlag Berlin Heidelberg 2011

in the basis $B_0 = \{x + y, x - y, 1; \text{sign}(x)\}$ are studied. Section 9.4 contains conclusions.

## 9.1 Bounds on Complexity of Decision Trees and Rules

In this section, we consider bounds on complexity of decision trees and decision rule systems over linear and quasilinear information systems.

### *9.1.1 Quasilinear Information Systems*

We will call a set $K$ a *numerical ring with unity* if $K \subseteq \mathbb{R}$, $1 \in K$, and for every $a, b \in K$ the relations $a + b \in K$, $a \times b \in K$ and $-a \in K$ hold. For instance, $\mathbb{R}$, $\mathbb{Q}$ and $\mathbb{Z}$ are numerical rings with unity.

Let $K$ be a numerical ring with unity, $A$ be a nonempty set and let $\varphi_1, \ldots, \varphi_m$ be functions from $A$ to $\mathbb{R}$. Denote

$$F(A, K, \varphi_1, \ldots, \varphi_m) = \left\{ \text{sign}\left( \sum_{i=1}^{m} d_i \varphi_i(x) + d_{m+1} \right) : d_1, \ldots, d_{m+1} \in K \right\} .$$

Here $\text{sign}(x) = -1$ if $x < 0$, $\text{sign}(x) = 0$ if $x = 0$, and $\text{sign}(x) = +1$ if $x > 0$. Set $E = \{-1, 0, +1\}$. The information system $(A, E, F(A, K, \varphi_1, \ldots, \varphi_m))$ will be denoted by $U(A, K, \varphi_1, \ldots, \varphi_m)$ and will be called a *quasilinear information system.*

Let $f \in F(A, K, \varphi_1, \ldots, \varphi_m)$ and $f = \text{sign}(\sum_{i=1}^{m} d_i \varphi_i(x) + d_{m+1})$. We define the parameter $r(f)$ of the attribute $f$ as follows. If $(d_1, \ldots, d_{m+1}) = (0, \ldots, 0)$ then $r(f) = 0$. Otherwise,

$$r(f) = \max\{0, \max\{\log_2 |d_i| : i \in \{1, \ldots, m + 1\}, d_i \neq 0\}\} .$$

For a problem $z = (\nu, f_1, \ldots, f_n)$ over $U(A, K, \varphi_1, \ldots, \varphi_m)$, set $r(z) = \max\{r(f_1), \ldots, r(f_n)\}$. Let $\Gamma$ be a decision tree over $U(A, K, \varphi_1, \ldots, \varphi_m)$ and $F(\Gamma)$ be the set of all attributes attached to nonterminal nodes of $\Gamma$. Denote $r(\Gamma) = \max\{r(f) : f \in F(\Gamma)\}$ (if $F(\Gamma) = \emptyset$ then $r(\Gamma) = 0$).

**Theorem 9.1.** ([42, 53]) *Let $U = U(A, K, \varphi_1, \ldots, \varphi_m)$ be a quasilinear information system. Then for each problem $z$ over $U$ there exists a decision tree $\Gamma$ over $U$ which solves the problem $z$ and for which the following inequalities hold:*

$$h(\Gamma) \leq (2(m + 2)^3 \log_2(\dim z + 2m + 2))/ \log_2(m + 2) ,$$

$$r(\Gamma) \leq 2(m + 1)^2(r(z) + 1 + \log_2(m + 1)) .$$

Let $U = U(A, K, \varphi_1, \ldots, \varphi_m)$ be a quasilinear information system. For a problem $z = (\nu, f_1, \ldots, f_n)$ over $U$, we denote by $h_U^g(z)$ the minimum depth of a decision tree over $U$ which solves the problem $z$. By $L_U^g(z)$ we denote the minimum depth of a decision rule system over $U$ which is complete for the problem $z$. We define two functions $h_U^g : \omega \setminus \{0\} \to \omega$ and $L_U^g : \omega \setminus \{0\} \to \omega$ in the following way:

$$h_U^g(n) = \max\{h_U^g(z) : z \in \mathcal{P}(U), \dim z \leq n\} ,$$
$$L_U^g(n) = \max\{L_U^g(z) : z \in \mathcal{P}(U), \dim z \leq n\}$$

for any $n \in \omega \setminus \{0\}$. The value $h_U^g(n)$ is the unimprovable upper bound on the value $h_U^g(z)$ for problems $z \in \mathcal{P}(U)$ such that $\dim z \leq n$. The value $L_U^g(n)$ is the unimprovable upper bound on the value $L_U^g(z)$ for problems $z \in \mathcal{P}(U)$ such that $\dim z \leq n$. The functions $h_U^g$ and $L_U^g$ are called *global Shannon functions* for the information system $U$.

The following theorem is a simple corollary of Theorems 8.9, 8.10 and 9.1.

**Theorem 9.2.** *Let $U = U(A, K, \varphi_1, \ldots, \varphi_m)$ be a quasilinear information system. Then the following statements hold:*

*a) if $\{(\varphi_1(a), \ldots, \varphi_m(a)) : a \in A\}$ is a finite set then $h_U^g(n) = O(1)$;*

*b) if $\{(\varphi_1(a), \ldots, \varphi_m(a)) : a \in A\}$ is an infinite set then $h_U^g(n) = \Theta(\log_2 n)$;*

*c) $L_U^g(n) = O(1)$.*

### 9.1.2 Linear Information Systems

A quasilinear information system $U(A, K, \varphi_1, \ldots, \varphi_m)$ is called *linear* if $A \subseteq \mathbb{R}^m$ and $\varphi_1 = x_1, \ldots, \varphi_m = x_m$. We will say about attributes from a linear information system as about *linear* attributes.

In this section, we consider a special kind of linear information system—the system $U_p^m = U(C_p^m, \mathbb{R}, x_1, \ldots, x_m)$ where $p$ is a positive real number and $C_p^m$ is the set of solutions on $\mathbb{R}^m$ of the equation system

$$\{-p < x_1 < p, \ldots, -p < x_m < p\} .$$

We will prove (based on some lemmas given without proofs) the following statement.

**Theorem 9.3.** *Let $m \geq 2$ and $p > 0$. Then for any problem $z \in \mathcal{P}(U_p^m)$ with $\dim z \geq 2$ there exist:*

*a) a decision tree $\Gamma$ over $U_p^m$ which solves $z$ and for which $h(\Gamma) \leq 2(m+1)^3 \log_2(\dim z + 2m)$;*

*b) a decision rule system $S$ over $U_p^m$ which is complete for $z$ and for which $L(S) \leq m + 1$.*

Bounds similar to the considered in part a) of Theorem 9.3 were obtained independently by two researchers and published in [36] and [33].

Denote $F_m = F(C_p^m, \mathbb{R}, x_1, \ldots, x_m)$ where

$$F(C_p^m, \mathrm{I\!R}, x_1, \ldots, x_m) = \left\{ \mathrm{sign}\left( \sum_{i=1}^{m} d_i x_i(x) + d_{m+1} \right) : d_1, \ldots, d_{m+1} \in \mathrm{I\!R} \right\} .$$

Let $V \subseteq C_p^m$ and $k$ be a natural number. A finite set $\mathcal{F} \subset F_m$ will be called a *k-functional cover for* $V$ if for any $\bar{a} \in V$ there exist attributes $f_1, \ldots, f_k \in \mathcal{F}$ and numbers $\delta_1, \ldots, \delta_k \in E = \{-1, 0, +1\}$ such that $\bar{a}$ is a solution of the equation system $\{f_1(\bar{x}) = \delta_1, \ldots, f_k(\bar{x}) = \delta_k\}$, and the set of solutions of this system on $\mathrm{I\!R}^m$ is a subset of $V$.

Consider three auxiliary statements.

**Lemma 9.4.** *Let $f_1, \ldots, f_n \in F_m$, $\delta_1, \ldots, \delta_n \in E$ and $V$ be the set of solutions from $C_p^m$ of the equation system $\{f_1(\bar{x}) = \delta_1, \ldots, f_n(\bar{x}) = \delta_n\}$. Then if $V$ is not empty then for $V$ there exists an $(m+1)$-functional cover $\mathcal{F}$ for which $|\mathcal{F}| \leq (m+1)(n+2m)^{m-1}$.*

**Lemma 9.5.** *Let $f_1, \ldots, f_n \in F_m$. Then the number of different tuples $(\delta_1, \ldots, \delta_n) \in E^n$ for each of which the system of equations $\{f_1(\bar{x}) = \delta_1, \ldots, f_n(\bar{x}) = \delta_n\}$ is compatible on $\mathrm{I\!R}^m$ (has a solution from $\mathrm{I\!R}^m$) is at most $2n^m + 1$.*

**Lemma 9.6.** *Any incompatible on $\mathrm{I\!R}^m$ system of equations of the kind*

$$\{f_1(\bar{x}) = \delta_1, \ldots, f_n(\bar{x}) = \delta_n\} ,$$

*where $f_1, \ldots, f_n \in F_m$ and $\delta_1, \ldots, \delta_n \in E$, contains an incompatible on $\mathrm{I\!R}^m$ subsystem with at most $m+1$ equations.*

*Proof (of Theorem 9.3).* Let $z = (\nu, f_1, \ldots, f_n) \in \mathcal{P}(U_p^m)$ and $n \geq 2$. Let the decision table $T(z)$ contain $k$ rows, and for $j = 1, \ldots, k$, let $(\delta_{j1}, \ldots, \delta_{jn})$ be the $j$-th row of $T(z)$. We denote by $W_j$ the set of solutions on $C_p^m$ of the equation system $\{f_1(\bar{x}) = \delta_{j1}, \ldots, f_n(\bar{x}) = \delta_{jn}\}$. From Lemma 9.4 it follows that for $W_j$ there exists an $(m+1)$-functional cover $\mathcal{F}_j$ for which $|\mathcal{F}_j| \leq (m+1)(n+2m)^{m-1}$.

We denote $\mathcal{F} = \bigcup_{j=1}^{k} \mathcal{F}_j \cup \{f_1, \ldots, f_n\}$. Let $r = |\mathcal{F}|$. Then $r \leq k(m+1)(n+2m)^{m-1} + n$. By Lemma 9.5, $k \leq 2n^m + 1$. Therefore $r \leq (2n^m + 1)(m+1)(n+2m)^{m-1} + n \leq (n+2m)^{2m} - 1$. Let $\mathcal{F} = \{g_1, \ldots, g_r\}$ where $f_1 = g_1, \ldots, f_n = g_n$. Let $\nu_1 : E^r \to \omega$ and for any $\bar{\delta} = (\delta_1, \ldots, \delta_r) \in E^r$ let $\nu_1(\bar{\delta}) = \nu(\delta_1, \ldots, \delta_n)$. Set $z_1 = (\nu_1, g_1, \ldots, g_r)$. From Lemma 9.5 it follows that $N(T(z_1)) \leq 2((n+2m)^{2m} - 1)^m + 1 \leq 2(n+2m)^{2m^2}$.

Let us show that $M(T(z_1)) \leq m+1$. Let $\bar{\delta} = (\delta_1, \ldots, \delta_r) \in E^r$. Let us show that there exist numbers $i_1, \ldots, i_t \in \{1, \ldots, r\}$ such that $t \leq m+1$ and the table $T(z_1)(g_{i_1}, \delta_{i_1}) \ldots (g_{i_t}, \delta_{i_t})$ is degenerate.

We consider two cases.

1) Let $\bar{\delta}$ be a row of $T(z_1)$ and $\bar{a} \in C_p^m$ be a solution of the equation system

$$\{g_1(\bar{x}) = \delta_1, \ldots, g_r(\bar{x}) = \delta_r\} . \tag{9.1}$$

Then for some $j \in \{1, \ldots, k\}$ we have $\bar{a} \in W_j$.

There exist functions $g_{i_1}, \ldots, g_{i_{m+1}} \in \mathcal{F}_j$ and numbers $\sigma_1, \ldots, \sigma_{m+1} \in E$ for which $\bar{a}$ is a solution of the equation system $\{g_{i_1}(\bar{x}) = \sigma_1, \ldots, g_{i_{m+1}}(\bar{x}) = \sigma_{m+1}\}$, and $W$—the set of solutions of this system on $\mathbb{R}^m$—is a subset of the set $W_j$. Taking into account that $g_{i_1}, \ldots, g_{i_{m+1}} \in \mathcal{F}_j$, we obtain $\sigma_1 = \delta_{i_1}, \ldots, \sigma_{m+1} = \delta_{i_{m+1}}$. The function $\nu_1(g_1(\bar{x}), \ldots, g_r(\bar{x}))$ is constant on $W_j$. Therefore this function is a constant on $W$. Hence

$$T(z_1)(g_{i_1}, \delta_{i_1}) \ldots (g_{i_{m+1}}, \delta_{i_{m+1}})$$

is a degenerate table and $M(T(z_1), \bar{\delta}) \leq m + 1$.

2) Let $\bar{\delta}$ be not a row of $T(z_1)$. Then the system of equations (9.1) is incompatible on $C_p^m$, and the system of equations

$$\begin{aligned} \{g_1(\bar{x}) = \delta_1, \ldots, g_r(\bar{x}) = \delta_r, \operatorname{sign}(x_1 - p) = -1, \\ \operatorname{sign}(x_1 + p) = +1, \ldots, \operatorname{sign}(x_m - p) = -1, \operatorname{sign}(x_m + p) = +1\} \end{aligned} \tag{9.2}$$

is incompatible on $\mathbb{R}^m$. From Lemma 9.6 it follows that there is a subsystem of the system (9.2) which is incompatible on $\mathbb{R}^m$ and has at most $m+1$ equations. Therefore there is a subsystem of the system (9.1) which is incompatible on $C_p^m$ and has at most $m+1$ equations. Let this system contain equations from (9.1) with numbers $i_1, \ldots, i_t$ where $t \leq m + 1$. Then $T(z_1)(g_{i_1}, \delta_{i_1}) \ldots (g_{i_t}, \delta_{i_t})$ is empty and therefore degenerate. So $M(T(z_1), \bar{\delta}) \leq t \leq m + 1$. Since $M(T(z_1), \bar{\delta}) \leq m + 1$ for any $\bar{\delta} \in E^r$, we have $M(T(z_1)) \leq m + 1$.

Using Theorem 3.17, we obtain

$$\begin{aligned} h(T(z_1)) &\leq M(T(z_1)) \log_2 N(T(z_1)) \leq (m+1) \log_2(2(n+2m)^{2m^2}) \\ &= (m+1)(2m^2 \log_2(n+2m) + 1) \leq 2(m+1)^3 \log_2(n+2m) . \end{aligned}$$

Therefore there exists a decision tree $\Gamma$ over $U_p^m$ which solves $z_1$ and for which $h(\Gamma) \leq 2(m+1)^3 \log_2(\dim z + 2m)$. By Theorem 3.11, $L(T(z_1)) \leq M(T(z_1)) \leq m + 1$. Therefore there exists a decision rule system $S$ over $U_p^m$ which is complete for $z_1$ and for which $L(S) \leq m + 1$. Evidently, $\nu_1(g_1(\bar{x}), \ldots, g_r(\bar{x})) \equiv \nu(f_1(\bar{x}), \ldots, f_n(\bar{x}))$. Therefore $\Gamma$ solves the problem $z$, and $S$ is complete for $z$. □

## 9.2 Optimization Problems over Quasilinear Information Systems

In this section, three classes of discrete optimization problems over quasilinear information systems are considered. For each class, examples and corollaries of Theorem 9.1 are given. More detailed discussion of considered results can be found in [42, 53].

### 9.2.1 Some Definitions

Let $U = U(A, K, \varphi_1, \ldots, \varphi_m)$ be a quasilinear information system. A pair $(A, \phi)$ where $\phi$ is a function from $A$ to a finite subset of the set $\omega$ will be called a *problem over A*. The problem $(A, \phi)$ may be interpreted as a problem of searching for the value $\phi(a)$ for a given $a \in A$. Let $k \in \omega$, $k \geq 1$, and $t \in \mathbb{R}$, $t \geq 0$. The problem $(A, \phi)$ will be called *$(m, k, t)$-problem over U* if there exists a problem $z$ over $U$ such that $\phi(a) = z(a)$ for each $a \in A$, $\dim z \leq k$ and $r(z) \leq t$. Let $\phi(a) = z(a)$ for each $a \in A$ and $z = (\nu, f_1, \ldots, f_p)$. Then the set $\{f_1, \ldots, f_p\}$ will be called a *separating set (with attributes from $F(A, K, \varphi_1, \ldots, \varphi_m)$)* for the problem $(A, \phi)$. We will say that a decision tree $\Gamma$ over $U$ *solves* the problem $(A, \phi)$ if the decision tree $\Gamma$ solves the problem $z$.

Denote

$$L(A, K, \varphi_1, \ldots, \varphi_m) = \left\{ \sum_{i=1}^{m} d_i \varphi_i(x) + d_{m+1} : d_1, \ldots, d_{m+1} \in K \right\} .$$

Let $g \in L(A, K, \varphi_1, \ldots, \varphi_m)$ and $g = \sum_{i=1}^{m} d_i \varphi_i(x) + d_{m+1}$. We define the parameter $r(g)$ of the function $g$ as follows. If $(d_1, \ldots, d_{m+1}) = (0, \ldots, 0)$ then $r(g) = 0$. Otherwise

$$r(g) = \max\{0, \max\{\log_2 |d_i| : i \in \{1, \ldots, m+1\}, d_i \neq 0\}\} .$$

In what follows we will assume that elements of the set $\{-1, 1\}^n$, of the set $\Pi_n$ of all $n$-degree permutations, and of the set $\{0, 1\}^n$ are enumerated by numbers from 1 to $2^n$, by numbers from 1 to $n!$, and by numbers from 1 to $2^n$ respectively.

### 9.2.2 Problems of Unconditional Optimization

Let $k \in \omega \setminus \{0\}$, $t \in \mathbb{R}, t \geq 0$, $g_1, \ldots, g_k \in L(A, K, \varphi_1, \ldots, \varphi_m)$, and $r(g_j) \leq t$ for $j = 1, \ldots, k$.

**Problem 9.7.** (Unconditional optimization of values of functions $g_1, \ldots, g_k$ on an element of the set $A$.) For a given $a \in A$ it is required to find the minimum number $i \in \{1, \ldots, k\}$ such that $g_i(a) = \min\{g_j(a) : 1 \leq j \leq k\}$.

One can show that the set $\{\text{sign}(g_i(x) - g_j(x)) : i, j \in \{1, \ldots, k\}, i \neq j\}$ is a separating set for this problem, and the considered problem is $(m, k^2, t+1)$-problem over the information system $U(A, K, \varphi_1, \ldots, \varphi_m)$.

*Example 9.8.* ($n$-City traveling salesman problem.) Let $n \in \omega$, $n \geq 4$, and let $G_n$ be the complete undirected graph with $n$ nodes. Assume that edges in $G_n$ are enumerated by numbers from 1 to $n(n-1)/2$, and Hamiltonian circuits in $G_n$ are enumerated by numbers from 1 to $(n-1)!/2$. Let a number

$a_i \in \mathbb{R}$ be attached to the $i$-th edge, $i = 1, \dots, n(n-1)/2$. We will interpret the number $a_i$ as the length of the $i$-th edge. It is required to find the minimum number of a Hamiltonian circuit in $G_n$ which has the minimum length. For each $j \in \{1, \dots, (n-1)!/2\}$, we will associate with the $j$-th Hamiltonian circuit the function $g_j(\bar{x}) = \sum_{i=1}^{n(n-1)/2} \delta_{ji} x_i$ where $\delta_{ji} = 1$ if the $i$-th edge is contained in the $j$-th Hamiltonian circuit, and $\delta_{ji} = 0$ otherwise. Obviously, the considered problem is the problem of unconditional optimization of values of functions $g_1, \dots, g_{(n-1)!/2}$ on an element of the set $\mathbb{R}^{n(n-1)/2}$. Therefore the set $\{\text{sign}(g_i(\bar{x}) - g_j(\bar{x})) : i, j \in \{1, \dots, (n-1)!/2\}, i \neq j\}$ is a separating set for the $n$-city traveling salesman problem, and this problem is $(n(n-1)/2, ((n-1)!/2)^2, 0)$-problem over the information system $U = U(\mathbb{R}^{n(n-1)/2}, \mathbb{Z}, x_1, \dots, x_{n(n-1)/2})$. From Theorem 9.1 it follows that there exists a decision tree $\Gamma$ over $U$ which solves the $n$-city traveling salesman problem and for which $h(\Gamma) \leq n^7/2$ and $r(\Gamma) \leq n^4 \log_2 n$.

*Example 9.9.* ($n$-Dimensional quadratic assignment problem.) Let $n \in \omega$ and $n \geq 2$. For given $a_{ij}, b_{ij} \in \mathbb{R}$, $1 \leq i, j \leq n$, it is required to find the minimum number of $n$-degree permutation $\pi$ which minimizes the value

$$\sum_{i=1}^{n} \sum_{j=1}^{n} a_{ij} b_{\pi(i)\pi(j)} .$$

Obviously, this problem is the problem of unconditional optimization of values of functions from the set $\{\sum_{i=1}^{n} \sum_{j=1}^{n} x_{ij} y_{\pi(i)\pi(j)} : \pi \in \Pi_n\}$ on an element of the set $\mathbb{R}^{2n^2}$. Hence the set

$$\{\text{sign}(\sum_{i=1}^{n} \sum_{j=1}^{n} x_{ij} y_{\pi(i)\pi(j)} - \sum_{i=1}^{n} \sum_{j=1}^{n} x_{ij} y_{\tau(i)\tau(j)}) : \pi, \tau \in \Pi_n, \pi \neq \tau\}$$

is a separating set for this problem, and the considered problem is $(n^4, (n!)^2, 0)$-problem over the information system $U = U(\mathbb{R}^{2n^2}, \mathbb{Z}, x_{11} y_{11}, \dots, x_{nn} y_{nn})$. From Theorem 9.1 it follows that there exists a decision tree $\Gamma$ over $U$ which solves the $n$-dimensional quadratic assignment problem and for which $h(\Gamma) \leq 3n(n^4 + 2)^3$ and $r(\Gamma) \leq 2(n^4 + 1)^2 \log_2(2n^4 + 2)$.

### 9.2.3 Problems of Unconditional Optimization of Absolute Values

Let $k \in \omega \setminus \{0\}$, $t \in \mathbb{R}, t \geq 0$, $g_1, \dots, g_k \in L(A, K, \varphi_1, \dots, \varphi_m)$, and $r(g_j) \leq t$ for $j = 1, \dots, k$.

**Problem 9.10.** (Unconditional optimization of absolute values of functions $g_1, \dots, g_k$ on an element of the set $A$.) For a given $a \in A$ it is required to find the minimum number $i \in \{1, \dots, k\}$ such that $|g_i(a)| = \min\{|g_j(a)| : 1 \leq j \leq k\}$.

Evidently, $|g_i(a)| < |g_j(a)|$ if and only if $(g_i(a)+g_j(a))(g_i(a)-g_j(a)) < 0$, and $|g_i(a)| = |g_j(a)|$ if and only if $(g_i(a) + g_j(a))(g_i(a) - g_j(a)) = 0$. Using these relations one can show that the set $\{\text{sign}(g_i(x) + g_j(x)), \text{sign}(g_i(x) - g_j(x)) : i, j \in \{1, \ldots, k\}, i \neq j\}$ is a separating set for the considered problem, and this problem is $(m, 2k^2, t+1)$-problem over the information system $U(A, K, \varphi_1, \ldots, \varphi_m)$.

*Example 9.11.* ($n$-Stone problem.) Let $n \in \omega \setminus \{0\}$. For a tuple $(a_1, \ldots, a_n) \in \mathbb{R}^n$ it is required to find the minimum number of a tuple $(\delta_1, \ldots, \delta_n) \in \{-1, 1\}^n$ which minimizes the value of $|\sum_{i=1}^n \delta_i a_i|$. Obviously, this problem is the problem of unconditional optimization of absolute values of functions from the set $\{\sum_{i=1}^n \delta_i x_i : (\delta_1, \ldots, \delta_n) \in \{-1, 1\}^n\}$ on an element of the set $\mathbb{R}^n$. Therefore the set $\{\text{sign}(\sum_{i=1}^n \delta_i x_i) : (\delta_1, \ldots, \delta_n) \in \{-2, 0, 2\}^n\}$ and, hence, the set $\{\text{sign}(\sum_{i=1}^n \delta_i x_i) : (\delta_1, \ldots, \delta_n) \in \{-1, 0, 1\}^n\}$ are separating sets for the considered problem, and this problem is $(n, 3^n, 0)$-problem over the information system $U = U(\mathbb{R}^n, \mathbb{Z}, x_1, \ldots, x_n)$. From Theorem 9.1 it follows that there exists a decision tree $\Gamma$ over $U$ which solves the $n$-stone problem and for which $h(\Gamma) \leq 4(n+2)^4/\log_2(n+2)$ and $r(\Gamma) \leq 2(n+1)^2 \log_2(2n+2)$.

### 9.2.4 Problems of Conditional Optimization

Let $k, p \in \omega \setminus \{0\}$, $t \in \mathbb{R}, t \geq 0$, $D \subseteq \mathbb{R}$, $D \neq \emptyset$ and $g_1, \ldots, g_k$ be functions from $L(A, K, \varphi_1, \ldots, \varphi_m)$ such that $r(g_j) \leq t$ for $j = 1, \ldots, k$.

**Problem 9.12.** (Conditional optimization of values of functions $g_1, \ldots, g_k$ on an element of the set $A$ with $p$ restrictions from $A \times D$.) For a given tuple $(a_0, a_1, \ldots, a_p, d_1, \ldots, d_p) \in A^{p+1} \times D^p$ it is required to find the minimum number $i \in \{1, \ldots, k\}$ such that $g_i(a_1) \leq d_1, \ldots, g_i(a_p) \leq d_p$ and $g_i(a_0) = \max\{g_j(a_0) : g_j(a_1) \leq d_1, \ldots, g_j(a_p) \leq d_p, j \in \{1, \ldots, k\}\}$ or to show that such $i$ does not exist. (In the last case let $k+1$ be the solution of the problem.)

The variables with values from $A$ will be denoted by $x_0, x_1, \ldots, x_p$ and the variables with values from $D$ will be denoted by $y_1, \ldots, y_p$. One can show that the set $\{\text{sign}(g_i(x_0) - g_j(x_0)) : 1 \leq i, j \leq k\} \cup (\bigcup_{j=1}^p \{\text{sign}(g_i(x_j) - y_j) : 1 \leq i \leq k\})$ is a separating set for the considered problem, and this problem is $(p + m(p+1), pk + k^2, t+1)$-problem over the information system $U(A^{p+1} \times D^p, K, \varphi_1(x_0), \ldots, \varphi_m(x_0), \ldots, \varphi_1(x_p), \ldots, \varphi_m(x_p), y_1, \ldots, y_p)$.

*Example 9.13.* (Problem on 0-1-knapsack with $n$ objects.) Let $n \in \omega \setminus \{0\}$. For a given tuple $(a_1, \ldots, a_{2n+1}) \in \mathbb{Z}^{2n+1}$, it is required to find the minimum number of a tuple $(\delta_1, \ldots, \delta_n) \in \{0, 1\}^n$ which maximizes the value $\sum_{i=1}^n \delta_i a_i$ under the condition $\sum_{i=1}^n \delta_i a_{n+i} \leq a_{2n+1}$. This is the problem of conditional optimization of values of functions from the set $\{\sum_{i=1}^n \delta_i x_i : (\delta_1, \ldots, \delta_n) \in \{0, 1\}^n\}$ on an element of the set $\mathbb{Z}^n$ with one restriction from $\mathbb{Z}^n \times \mathbb{Z}$. The set $\{\text{sign}(\sum_{i=1}^n \delta_i x_i) : (\delta_1, \ldots, \delta_n) \in \{-1, 0, 1\}^n\} \cup \{\text{sign}(\sum_{i=1}^n \delta_i x_{n+i} -$

$x_{2n+1}) : (\delta_1, \ldots, \delta_n) \in \{0,1\}^n\}$ is a separating set for the considered problem, and this problem is $(2n+1, 3^n + 2^n, 0)$-problem over the information system $U = U(\mathbb{Z}^{2n+1}, \mathbb{Z}, x_1, \ldots, x_{2n+1})$. From Theorem 9.1 it follows that there exists a decision tree $\Gamma$ over $U$ which solves the problem on 0-1-knapsack with $n$ objects and for which $h(\Gamma) \leq 2(2n+3)^4 / \log_2(2n+3)$ and $r(\Gamma) \leq 2(2n+2)^2 \log_2(4n+4)$.

## 9.3 On Depth of Acyclic Programs

In this section, relationships between depth of deterministic and nondeterministic acyclic programs in the basis $B_0 = \{x+y, x-y, 1; \mathrm{sign}(x)\}$ are considered. Proof of the main result of this section (see [40, 53]) is based on Theorem 9.1 and is an example of the application of methods of decision tree theory to analysis of algorithms which are not decision trees.

### *9.3.1 Main Definitions*

Letters from the alphabet $X = \{x_i : i \in \omega\}$ will be called *input variables*, while letters from the alphabet $Y = \{y_i : i \in \omega\}$ will be called *working variables*.

A *program in the basis* $B_0$ is a labeled finite directed graph which has nodes of the following six kinds:

a) the only node without entering edges called the node "input";

b) the only node without issuing edges called the node "output";

c) functional nodes of the kinds $y_j := 1, y_j := z_l + z_k$ and $y_j := z_l - z_k$ where $z_l, z_k \in X \cup Y$;

d) predicate nodes of the kind $\mathrm{sign}(y_j)$.

Each edge issuing from a predicate node is labeled with a number from the set $\{-1, 0, +1\}$. The other edges are not labeled.

Further we assume that in expressions assigned to nodes of a program there is at least one input and, hence, at least one working variable.

A program in the basis $B_0$ will be called *acyclic* if it contains no directed cycles. A program will be called *deterministic* if it satisfies the following conditions: the node "input" and each functional node have exactly one issuing edge, and edges issuing from a predicate node are assigned pairwise different numbers. If a program isn't deterministic we will call it *nondeterministic*.

Let $P$ be an acyclic program in the basis $B_0$ with the input variables $x_1, \ldots, x_n$ and the working variables $y_1, \ldots, y_t$.

A *complete path* in $P$ is an arbitrary directed path from the node "input" to the node "output". Let $\xi = v_1, d_1, \ldots, v_m, d_m, v_{m+1}$ be a complete path in the program $P$. Define the set of elements from $\mathbb{Q}^n$ accepted by the complete path $\xi$. For $i = 1, \ldots, m$, we will attach to the node $v_i$ of the path $\xi$ a tuple $\bar{\beta}_i = (\beta_{i1}, \ldots, \beta_{it})$ composed of functions from the set $\{\sum_{i=1}^{n} b_i x_i + b_{n+1} : b_1, \ldots, b_{n+1} \in \mathbb{Z}\}$. Let $\bar{\beta}_1 = (0, \ldots, 0)$. Let the tuples $\bar{\beta}_1, \ldots, \bar{\beta}_{i-1}$, where

$2 \leq i \leq m$, be already defined. If $v_i$ is a predicate node then $\bar{\beta}_i = \bar{\beta}_{i-1}$. Let $v_i$ be a functional node and let for definiteness the node $v_i$ be of the kind $y_j := x_l + y_p$. Then $\bar{\beta}_i = (\beta_{i-11}, \ldots, \beta_{i-1j-1}, x_l + \beta_{i-1p}, \beta_{i-1j+1}, \ldots, \beta_{i-1t})$. For other kinds of functional nodes, the tuple $\bar{\beta}_i$ is defined in the same way.

Let $v_{i_1}, \ldots, v_{i_k}$ be all predicate nodes in the complete path $\xi$. Let $k > 0$, the nodes $v_{i_1}, \ldots, v_{i_k}$ be of the kind $\text{sign}(y_{j_1}), \ldots, \text{sign}(y_{j_k})$, and the edges $d_{i_1}, \ldots, d_{i_k}$ be labeled with the numbers $\delta_1, \ldots, \delta_k$. Denote $\mathcal{A}(\xi)$ the set of solutions on $\mathbb{Q}^n$ of the equation system

$$\{\text{sign}(\beta_{i_1 j_1}(\bar{x})) = \delta_1, \ldots, \text{sign}(\beta_{i_k j_k}(\bar{x})) = \delta_k\} .$$

If $k = 0$ then $\mathcal{A}(\xi) = \mathbb{Q}^n$. The set $\mathcal{A}(\xi)$ will be called the *set of elements from* $\mathbb{Q}^n$ *accepted by the complete path* $\xi$. The set of all complete paths in the program $P$ will be denoted by $\Xi(P)$. Evidently, $\Xi(P) \neq \emptyset$. Denote $\mathcal{A}(P) = \bigcup_{\xi \in \Xi(P)} \mathcal{A}(\xi)$. We will say that the *program* $P$ *recognizes the set* $A(P)$.

Denote by $h(\xi)$ the number of functional and predicate nodes in a complete path $\xi$. The value $h(P) = \max\{h(\xi) : \xi \in \Xi(P)\}$ will be called the *depth* of the program $P$.

### 9.3.2 Relationships between Depth of Deterministic and Nondeterministic Acyclic Programs

Acyclic programs $P_1$ and $P_2$ in the basis $B_0$ will be called *equivalent* if the sets of input variables of $P_1$ and $P_2$ coincide, and the equality $\mathcal{A}(P_1) = \mathcal{A}(P_2)$ holds.

**Theorem 9.14.** ([40, 53]) *For each nondeterministic acyclic program $P_1$ in the basis $B_0$ with $n$ input variables there exists a deterministic acyclic program $P_2$ in the basis $B_0$ which is equivalent to $P_1$ and for which the following inequality holds:*

$$h(P_2) \leq 8(n+2)^7(h(P_1)+2)^2 .$$

Analogous upper bound was obtained in [34] for simulation of parallel acyclic programs in similar basis by decision trees.

*Example 9.15.* (Problem of partition of $n$ numbers.) We denote by $W_n$ the set of tuples $(q_1, \ldots, q_n)$ from $\mathbb{Q}^n$ for each of which there exists a tuple $(\sigma_1, \ldots, \sigma_n) \in \{-1, 1\}^n$ such that $\sum_{i=1}^{n} \sigma_i q_i = 0$. The problem of recognition of belonging a tuple from $\mathbb{Q}^n$ to the set $W_n$ is known as the problem of partition of $n$ numbers. Figure 9.1 represents a nondeterministic acyclic program $P_n$ in the basis $B_0$ with input variables $x_1, \ldots, x_n$ and working variable $y_1$ for which $\mathcal{A}(P_n) = W_n$ and $h(P_n) = n + 1$. From Theorem 9.14 it follows that there exists a deterministic acyclic program in the basis $B_0$ which recognizes the set $W_n$ and for which the depth is at most $8(n+3)^9$.

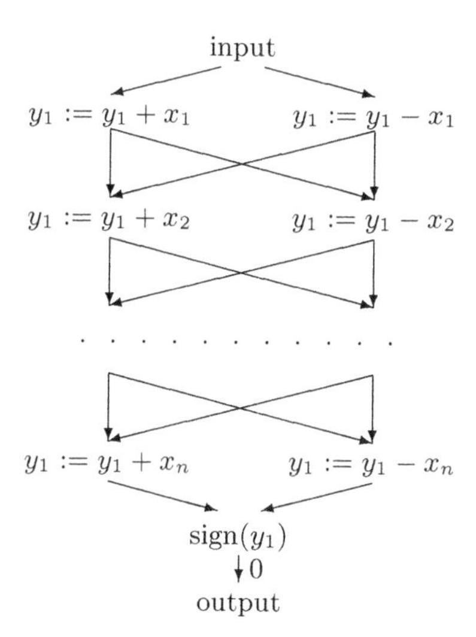


**Fig. 9.1**

## 9.4 Conclusions

This chapter contains bounds (mainly without proofs) on time complexity of decision trees and decision rule systems over linear and quasilinear information systems, and applications of discussed results to discrete optimization problems and analysis of acyclic programs in the basis $\{x+y, x-y, 1; \mathrm{sign}(x)\}$.

We proved that for some known problems of discrete optimization (including a number of $NP$-hard problems) there exist decision trees with small depth. In particular, we proved that there exists a decision tree $\Gamma$ with linear attributes which solves the $n$-stone problem and for which the depth is at most $4(n+2)^4$. There exists a decision tree $\Gamma$ with linear attributes which solves the $n$-city traveling salesman problem and for which the depth is at most $n^7$. The question about possibilities of efficient use of such trees is open. However, we can prove that the number of nodes in such trees is large.

In particular, in [35] (see also [49]) it was proved that the minimum cardinality of separating set with attributes from $F(\mathbb{R}^n, \mathbb{Z}, x_1, \ldots, x_n)$ for the $n$-stone problem is equal to $(3^n - 2^n - 1)/2$. Using this fact one can show that any decision tree with attributes from $F(\mathbb{R}^n, \mathbb{Z}, x_1, \ldots, x_n)$ which solves the $n$-stone problem has at least $(3^n - 2^n - 1)/2$ working nodes. This result can be extended to decision diagrams (branching programs).

# 10

# Recognition of Words and Diagnosis of Faults

In this chapter, we consider two more applications: recognition of regular language words and diagnosis of constant faults in combinatorial circuits. In the first case, we study both decision trees and complete systems of decision rules. In the second case, we restrict our consideration to decision trees.

Proofs are too complicated to be considered in this chapter. However, we give some comments relative to the tools used in the proofs or ideas of proofs.

This chapter consists of three sections. In Sect. 10.1, the problem of regular language word recognition is studied. Section 10.2 is devoted to the problem of diagnosis of constant faults in combinatorial circuits. Section 10.3 contains conclusions.

## 10.1 Regular Language Word Recognition

In this section, we consider the problem of recognition of words of fixed length in a regular language. The word under consideration can be interpreted as a description of certain screen image in the following way: the $i$-th letter of the word encodes the color of the $i$-th screen cell. In this case, a decision tree (or a decision rule system) which recognizes some words may be interpreted as an algorithm for the recognition of images which are defined by these words. The considered here results (mainly with proofs) can be found in [46, 53].

### *10.1.1 Problem of Recognition of Words*

Let $k \in \omega$, $k \geq 2$ and $E_k = \{0, 1, \ldots, k-1\}$. By $(E_k)^*$ we denote the set of all finite words over the alphabet $E_k$, including the empty word $\lambda$. Let $\mathcal{L}$ be a regular language over the alphabet $E_k$. For $n \in \omega \setminus \{0\}$, we denote by $\mathcal{L}(n)$ the set of all words from $\mathcal{L}$ for which the length is equal to $n$. Let us assume that $\mathcal{L}(n) \neq \emptyset$. For $i \in \{1, \ldots, n\}$, we define a function $l_i : \mathcal{L}(n) \to E_k$ as follows: $l_i(\delta_1 \ldots \delta_n) = \delta_i$ for each $\delta_1 \ldots \delta_n \in \mathcal{L}(n)$. Let us consider an information system $U(\mathcal{L}, n) = (\mathcal{L}(n), E_k, \{l_1, \ldots, l_n\})$ and a problem $z_{\mathcal{L},n} = (\nu, l_1, \ldots, l_n)$

M. Moshkov and B. Zielosko: Combinatorial Machine Learning, SCI 360, pp. 155–170.
springerlink.com © Springer-Verlag Berlin Heidelberg 2011

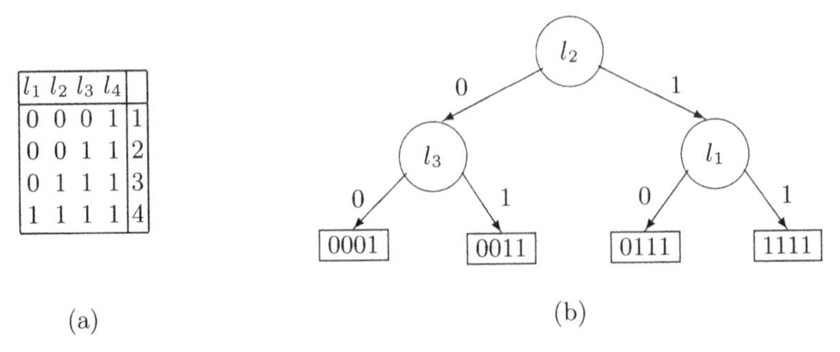


**Fig. 10.1**

over $U(\mathcal{L}, n)$ such that $\nu(\bar{\delta}_1) \neq \nu(\bar{\delta}_2)$ for every $\bar{\delta}_1, \bar{\delta}_2 \in E_k^n$, $\bar{\delta}_1 \neq \bar{\delta}_2$. The problem $z_{\mathcal{L},n}$ will be called the *problem of recognition of words from* $\mathcal{L}(n)$.

We denote by $h_{\mathcal{L}}(n)$ the minimum depth of a decision tree over $U(\mathcal{L}, n)$ which solves the problem of recognition of words from $\mathcal{L}(n)$. If $\mathcal{L}(n) = \emptyset$ then $h_{\mathcal{L}}(n) = 0$. We denote by $L_{\mathcal{L}}(n)$ the minimum depth of a decision rule system over $U(\mathcal{L}, n)$ which is complete for the problem of recognition of words from $\mathcal{L}(n)$. If $\mathcal{L}(n) = \emptyset$ then $L_{\mathcal{L}}(n) = 0$.

In this section, we consider the behavior of two functions $H_{\mathcal{L}} : \omega \setminus \{0\} \to \omega$ and $P_{\mathcal{L}} : \omega \setminus \{0\} \to \omega$ which are defined as follows. Let $n \in \omega \setminus \{0\}$. Then

$$H_{\mathcal{L}}(n) = \max\{h_{\mathcal{L}}(m) : m \in \omega \setminus \{0\}, m \leq n\} ,$$
$$P_{\mathcal{L}}(n) = \max\{L_{\mathcal{L}}(m) : m \in \omega \setminus \{0\}, m \leq n\} .$$

*Example 10.1.* Let $\mathcal{L}$ be the regular language which is generated by the source represented in Fig. 10.3. Let us consider the problem $z_{\mathcal{L},4} = (\nu, l_1, l_2, l_3, l_4)$ of recognition of words from $\mathcal{L}(4) = \{0001, 0011, 0111, 1111\}$. Let $\nu(0,0,0,1) = 1$, $\nu(0,0,1,1) = 2$, $\nu(0,1,1,1) = 3$ and $\nu(1,1,1,1) = 4$. The decision table $T(z_{\mathcal{L},4})$ is represented in Fig. 10.1(a). The decision tree in Fig. 10.1(b) solves the problem of recognition of words from $\mathcal{L}(4)$. Note that instead of numbers of words the terminal nodes in this tree are labeled with words. The depth of the considered decision tree is equal to 2. Using Theorem 3.1, we obtain $h_{\mathcal{L}}(4) = 2$. The decision rule system

$$\{l_3 = 0 \to 1, l_2 = 0 \wedge l_3 = 1 \to 2, l_1 = 0 \wedge l_2 = 1 \to 3, l_1 = 1 \to 4\}$$

is complete for the problem of recognition of words from $\mathcal{L}(4)$. The depth of this system is equal to 2. One can show that $M(T(z_{\mathcal{L},4}), (0,0,1,1)) = 2$. Using Theorem 3.11, we obtain $L_{\mathcal{L}}(4) = 2$.

### 10.1.2 A-Sources

An *A-source over the alphabet* $E_k$ is a triple $I = (G, q_0, Q)$ where $G$ is a directed graph, possibly with multiple edges and loops, in which each edge is labeled with a number from $E_k$ and any edges starting in a node are labeled with pairwise different numbers, $q_0$ is a node of $G$, and $Q$ is a nonempty set of the graph $G$ nodes.

Let $I = (G, q_0, Q)$ be an A-source over the alphabet $E_k$. An *I-trace* in the graph $G$ is an arbitrary sequence $\tau = v_1, d_1, \ldots, v_m, d_m, v_{m+1}$ of nodes and edges of $G$ such that $v_1 = q_0$, $v_{m+1} \in Q$, and $v_i$ is the initial and $v_{i+1}$ is the terminal node of the edge $d_i$ for $i = 1, \ldots, m$. Now we define a word word$(\tau)$ from $(E_k)^*$ in the following way: if $m = 0$ then word$(\tau) = \lambda$. Let $m > 0$, and let $\delta_j$ be the number assigned to the edge $d_j$, $j = 1, \ldots, m$. Then word$(\tau) = \delta_1 \ldots \delta_m$. We can extend the notation word$(\tau)$ to an arbitrary directed path $\tau$ in the graph $G$. We denote by $\Xi(I)$ the set of all $I$-traces in $G$. Let $E(I) = \{\text{word}(\tau) : \tau \in \Xi(I)\}$. We will say that the source $I$ *generates* the language $E(I)$. It is well known that $E(I)$ is a regular language.

The A-source $I$ will be called *everywhere defined over the alphabet* $E_k$ if each node of $G$ is the initial node of exactly $k$ edges which are assigned pairwise different numbers from $E_k$. The A-source $I$ will be called *reduced* if for each node of $G$ there exists an $I$-trace which contains this node. It is known [32] that for each regular language over the alphabet $E_k$ there exists an everywhere defined over the alphabet $E_k$ A-source which generates this language. Therefore for each nonempty regular language there exists a reduced A-source which generates this language. Further we will assume that a considered regular language is nonempty and it is given by a reduced A-source which generates this language.

### 10.1.3 Types of Reduced A-Sources

Let $I = (G, q_0, Q)$ be a reduced A-source over the alphabet $E_k$. A directed path in the graph $G$ will be called a *path of the source* $I$. A path of the source $I$ will be called a *cycle of the source* $I$ if there is at least one edge in this path, and the first node of this path is equal to the last node of this path. A cycle of the source $I$ will be called *elementary* if nodes of this cycle, with the exception of the last node, are pairwise different. Sometimes, cycles of the source $I$ will be considered as subgraphs of the graph $G$.

The source $I$ will be called *simple* if each two different (as subgraphs) elementary cycles of the source $I$ do not have common nodes. Let $I$ be a simple source and $\tau$ be an $I$-trace. The number of different (as subgraphs) elementary cycles of the source $I$, which have common nodes with $\tau$, will be denoted by $cl(\tau)$ and will be called the *cyclic length of the path* $\tau$. The value $cl(I) = \max\{cl(\tau) : \tau \in \Xi(I)\}$ will be called the *cyclic length of the source* $I$.

Let $I$ be a simple source, $C$ be an elementary cycle of the source $I$, and $v$ be a node of the cycle $C$. Beginning with the node $v$, the cycle $C$ generates an infinite periodic word over the alphabet $E_k$. This word will be denoted by $W(I, C, v)$. We denote by $r(I, C, v)$ the minimum period of the word $W(I, C, v)$. We denote by $l(C)$ the number of nodes in the elementary cycle $C$ (the *length* of $C$).

The source $I$ will be called *dependent* if there exist two different (as subgraphs) elementary cycles $C_1$ and $C_2$ of the source $I$, nodes $v_1$ and $v_2$ of the cycles $C_1$ and $C_2$ respectively, and a path $\pi$ of the source $I$ from $v_1$ to $v_2$ which satisfy the following conditions: $W(I, C_1, v_1) = W(I, C_2, v_2)$ and the length of the path $\pi$ is a number divisible by $r(I, C_1, v_1)$. If the source $I$ is not dependent then it will be called *independent*.

The source $I$ will be called *strongly dependent* if in $I$ there exist pairwise different (as subgraphs) elementary cycles $C_1, \ldots, C_m$, $(m \geq 1)$, and pairwise different (as subgraphs) elementary cycles $B_1, \ldots, B_m$, $D_1, \ldots, D_m$, vertices $v_0, \ldots, v_{m+1}, u_1, \ldots, v_{m+1}, w_0, \ldots, w_m$ and paths $\tau_0, \ldots, \tau_m, \pi_0, \ldots, \pi_m$, $\gamma_1, \ldots, \gamma_m$ which satisfy the following conditions:

a) $v_0 = w_0 = q_0$, $v_{m+1} \in Q$, $u_{m+1} \in Q$;
b) for $i = 1, \ldots, m$, the node $v_i$ belongs to the cycle $C_i$, the node $u_i$ belongs to the cycle $B_i$, and the node $w_i$ belongs to the cycle $D_i$;
c) $\tau_i$ is a path from $v_i$ to $v_{i+1}$, and $\pi_i$ is a path from $w_i$ to $u_{i+1}$, $i = 0, \ldots, m$;
d) $\gamma_i$ is a path from $u_i$ to $w_i$, $i = 1, \ldots, m$;
e) $W(I, C_i, v_i) = W(I, B_i, u_i) = W(I, D_i, w_i)$ for $i = 1, \ldots, m$;
f) $\text{word}(\tau_i) = \text{word}(\pi_i)$ for $i = 0, \ldots, m$;
g) for $i = 1, \ldots, m$, the length of the path $\gamma_i$ is a number divisible by $l(C_i)$.

One can show that if the source $I$ is strongly dependent then the source $I$ is dependent.

### *10.1.4 Main Result*

In the following theorem, the behavior of functions $H_{\mathcal{L}}$ and $P_{\mathcal{L}}$ is considered.

**Theorem 10.2.** *Let $\mathcal{L}$ be a nonempty regular language and $I$ be a reduced $A$-source which generates the language $\mathcal{L}$. Then*

*a) if $I$ is an independent simple source and $cl(I) \leq 1$ then there exists a constant $c_1 \in \omega \setminus \{0\}$ such that for any $n \in \omega \setminus \{0\}$ the following inequalities hold:*

$$P_{\mathcal{L}}(n) \leq H_{\mathcal{L}}(n) \leq c_1 ;$$

*b) if $I$ is an independent simple source and $cl(I) \geq 2$ then there exist constants $c_1, c_2, c_3, c_4, c_5 \in \omega \setminus \{0\}$ such that for any $n \in \omega \setminus \{0\}$ the following inequalities hold:*

$$\frac{\log_2 n}{c_1} - c_2 \leq H_{\mathcal{L}}(n) \leq c_3 \log_2 n + c_4 \ \ \text{and} \ \ P_{\mathcal{L}}(n) \leq c_5 ;$$

*c) if $I$ is a dependent simple source which is not strongly dependent then there exist constants $c_1, c_2, c_3 \in \omega \setminus \{0\}$ such that for any $n \in \omega \setminus \{0\}$ the following inequalities hold:*

$$\frac{n}{c_1} - c_2 \leq H_{\mathcal{L}}(n) \leq n \text{ and } P_{\mathcal{L}}(n) \leq c_3 ;$$

*d) if $I$ is a strongly dependent simple source or $I$ is not a simple source then there exist constants $c_1, c_2 \in \omega \setminus \{0\}$ such that for any $n \in \omega \setminus \{0\}$ the following inequalities hold:*

$$\frac{n}{c_1} - c_2 \leq P_{\mathcal{L}}(n) \leq H_{\mathcal{L}}(n) \leq n .$$

Proof of this theorem is too long and complicated to be considered in this book. However, we can give some explanations regarding tools used in this proof. There are Theorem 3.1 (generalized to the case of $k$-valued decision tables), Theorems 3.6, 3.11, 3.14, 3.17 and Corollary 2.24.

Let us consider, for example, the bound $H_{\mathcal{L}}(n) \leq c_3 \log_2 n + c_4$ for the case when $I$ is an independent simple source and $cl(I) \geq 2$. One can show that in this case there exist two constants $c_1, c_2 \in \omega \setminus \{0\}$ such that for any $n \in \omega \setminus \{0\}$ the following inequalities hold: $N(T(z_{\mathcal{L},n})) \leq c_1 n^{cl(I)}$ and $M(T(z_{\mathcal{L},n})) \leq c_2$. Using Theorem 3.17, we obtain

$$\begin{aligned} h_{\mathcal{L}}(n) &= h(T(z_{\mathcal{L},n})) \leq M(T(z_{\mathcal{L},n})) \log_2 N(T(z_{\mathcal{L},n})) \\ &= c_2 cl(I)(\log_2 c_1 + \log_2 n) \leq c_3 \log_2 n + c_4 , \end{aligned}$$

where $c_3 = c_2 cl(I)$ and $c_4 = c_2 cl(I) \lceil \log_2 c_1 \rceil + 1$. Thus $H_{\mathcal{L}}(n) \leq c_3 \log_2 n + c_4$.

### 10.1.5 Examples

Further in figures for a source $I = (G, q_0, Q)$ the node $q_0$ will be labeled with $+$, and each node from $Q$ will be labeled with $*$.

*Example 10.3.* Let $I_1$ be the source depicted in Fig. 10.2 and $\mathcal{L}_1$ be the regular language which is generated by $I_1$. The source $I_1$ is an independent simple A-source with $cl(I_1) = 1$. One can show that $P_{\mathcal{L}_1}(n) = H_{\mathcal{L}_1}(n) = 0$ for any $n \in \omega \setminus \{0\}$.

*Example 10.4.* Let $I_2$ be the source depicted in Fig. 10.3 and $\mathcal{L}_2$ be the regular language which is generated by $I_2$. The source $I_2$ is an independent simple A-source with $cl(I_2) = 2$. One can show that $H_{\mathcal{L}_2}(n) = \lceil \log_2 n \rceil$ for any $n \in \omega \setminus \{0\}$, $P_{\mathcal{L}_2}(n) = n - 1$ for $n = 1, 2$ and $P_{\mathcal{L}_2}(n) = 2$ for any $n \in \omega \setminus \{0\}, n \geq 3$.

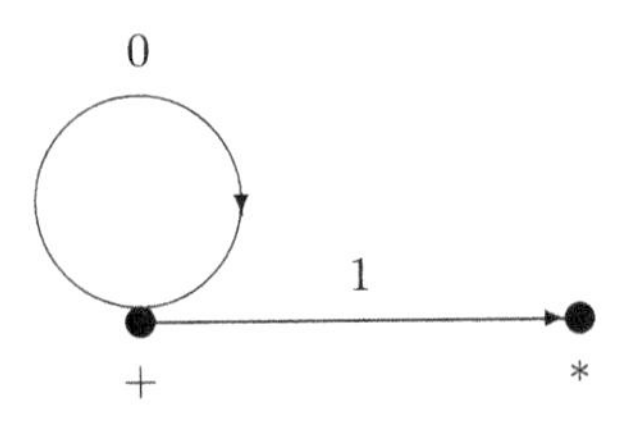


**Fig. 10.2**

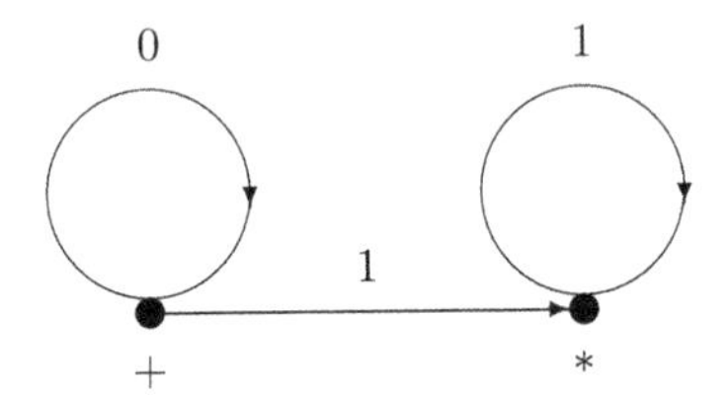


**Fig. 10.3**

*Example 10.5.* Let $I_3$ be the source depicted in Fig. 10.4 and $\mathcal{L}_3$ be the regular language which is generated by $I_3$. The source $I_3$ is a dependent simple A-source which is not strongly dependent. One can show that $H_{\mathcal{L}_3}(n) = n - 1$ for any $n \in \omega \setminus \{0\}$, $P_{\mathcal{L}_3}(1) = 0$, and $P_{\mathcal{L}_3}(n) = 1$ for any $n \in \omega \setminus \{0, 1\}$.

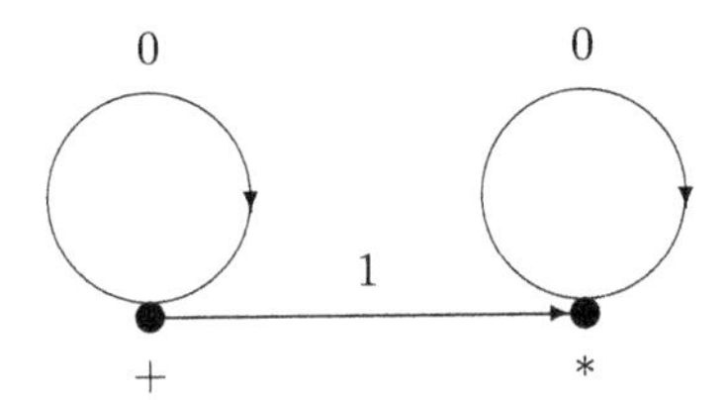


**Fig. 10.4**

*Example 10.6.* Let $I_4$ be the source depicted in Fig. 10.5 and $\mathcal{L}_4$ be the regular language which is generated by $I_4$. The source $I_4$ is a strongly dependent simple A-source. One can show that $H_{\mathcal{L}_4}(n) = P_{\mathcal{L}_4}(n) = n$ for any $n \in \omega \setminus \{0\}$.

*Example 10.7.* Let $I_5$ be the source depicted in Fig. 10.6 and $\mathcal{L}_5$ be the regular language which is generated by $I_5$. The source $I_5$ is an A-source which is not simple. One can show that $H_{\mathcal{L}_5}(n) = P_{\mathcal{L}_5}(n) = n$ for any $n \in \omega \setminus \{0\}$.

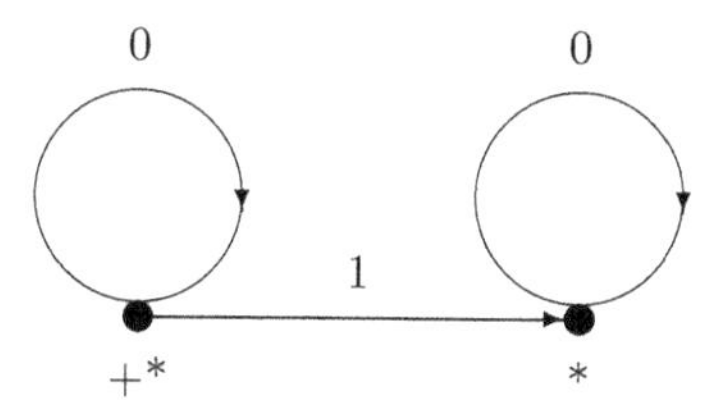


**Fig. 10.5**

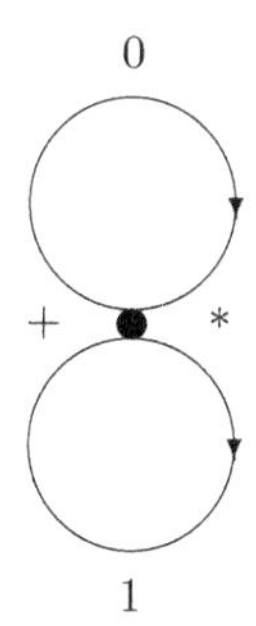


**Fig. 10.6**

## 10.2 Diagnosis of Constant Faults in Circuits

The problem of constant fault diagnosis in combinatorial circuits is studied in this section. Faults under consideration are represented in the form of Boolean constants on some inputs of the circuit gates. The diagnosis problem consists in the recognition of the function realized by the circuit with a fixed tuple of constant faults from given set of tuples. For this problem solving we use decision trees. Each attribute in a decision tree consists in observation of output of the circuit at the inputs of which a binary tuple is given.

As for the problem of regular language word recognition, proofs are too complicated to be considered here. The most part of lower bounds is based on Theorem 3.6. Upper bounds are based on specific algorithms of diagnosis. We explain the ideas of these algorithms. The results with proofs can be found in [48, 53].

### *10.2.1 Basic Notions*

The notions of combinatorial circuit, set of tuples of constant faults and diagnosis problem are defined in this subsection.

## Combinatorial Circuits

A *basis* is an arbitrary nonempty finite set of Boolean functions. Let $B$ be a basis.

A *combinatorial circuit in the basis B* (a *circuit in the basis B*) is a labeled finite directed acyclic graph with multiple edges which has nodes of the three types: inputs, gates and outputs.

Nodes of the *input* type have no entering edges, each input is labeled with a variable, and distinct inputs are labeled with distinct variables. Every circuit has at least one input.

Each node of the *gate* type is labeled with a function from the set $B$. Let $v$ be a gate and let a function $g$ depending on $t$ variables be attached to it. If $t = 0$ (this is the case when $g$ is one of the constants 0 or 1) then the node $v$ has no entering edges. If $t > 0$ then the node $v$ has exactly $t$ entering edges which are labeled with numbers $1, \ldots, t$ respectively. Every circuit has at least one gate.

Each node of the *output* type has exactly one entering edge which issues from a gate. Let $v$ be an output. Nothing is attached to $v$, and $v$ has no issuing edges. We will consider only circuits which have exactly one output.

Let $S$ be a circuit in the basis $B$ which has $n$ inputs labeled with variables $x_1, \ldots, x_n$. Let us correspond to each node $v$ in the circuit $S$ a Boolean function $f_v$ depending on variables $x_1, \ldots, x_n$. If $v$ is an input of $S$ labeled with the variable $x_i$ then $f_v = x_i$. If $v$ is a gate labeled with a constant $c \in \{0, 1\}$ then $f_v = c$. Let $v$ be a gate labeled with a function $g$ depending on $t > 0$ variables. For $i = 1, \ldots, t$, let the edge $d_i$, labeled with the number $i$, issue from a node $v_i$ and enter the node $v$. Then $f_v = g(f_{v_1}, \ldots, f_{v_t})$. If $v$ is an output of the circuit $S$ and an edge, issuing from a node $u$, enters the node $v$, then $f_v = f_u$. The function corresponding to the output of the circuit $S$ will be denoted by $f_S$. We will say that the circuit $S$ *realizes* the function $f_S$.

Denote by $\#(S)$ the number of gates in the circuit $S$. The value $\#(S)$ characterizes the complexity of the circuit $S$.

Denote by $\mathrm{Circ}(B)$ the set of circuits in the basis $B$ and by $\mathcal{F}(B)$—the set $\{f_S : S \in \mathrm{Circ}(B)\}$ of functions realized by circuits in the basis $B$.

## Set of Tuples of Constant Faults on Inputs of Gates

Let $S$ be a circuit in basis $B$. Edges entering gates of the circuit $S$ will be called *inputs of gates*. Let the circuit $S$ have $m$ gate inputs. The circuit $S$ will be called *degenerate* if $m = 0$ and *nondegenerate* if $m > 0$. Let $S$ be a nondegenerate circuit. Later we will assume that the gate inputs in the circuit $S$ are enumerated by numbers from 1 to $m$. Thus, each edge entering a gate has a sequential number in the circuit besides the number attached to it and corresponding to the gate.

We will consider the faults in the circuit $S$ which consist in appearance of Boolean constants on gate inputs. Each fault of such kind is defined by a *tuple of constant faults on inputs of gates of the circuit* $S$ which is an arbitrary $m$-tuple of the kind $\bar{w} = (w_1, \ldots, w_m) \in \{0,1,2\}^m$. If $w_i = 2$ then the $i$-th gate input in the circuit $S$ operates properly. If $w_i \neq 2$ then the $i$-th gate input in the circuit $S$ is faulty and realizes the constant $w_i$.

Define a circuit $S(\bar{w})$ in the basis $B \cup \{0,1\}$ which will be interpreted as the result of action of the tuple of faults $\bar{w}$ on the circuit $S$. Let us overlook all gate inputs in the circuit $S$. Let $i \in \{1, \ldots, m\}$. If $w_i = 2$ then the $i$-th gate input will be left without changes. Let $w_i \neq 2$ and the $i$-th gate input is the edge $d$ issuing from the node $v_1$ and entering the node $v_2$. Add to the circuit $S$ new gate $v(w_i)$ which is labeled with the constant $w_i$. Instead of the node $v_1$ connect the edge $d$ to the node $v(w_i)$.

A *set of tuples of constant faults on inputs of gates of the circuit* $S$ is a subset $W$ of the set $\{0,1,2\}^m$ containing the tuple $(2, \ldots, 2)$. Denote $\mathrm{Circ}(S, W) = \{S(\bar{w}) : \bar{w} \in W\}$. Note that $S((2, \ldots, 2)) = S$.

## Problem of Diagnosis

Let $S$ be a nondegenerate circuit in the basis $B$ with $n$ inputs and $m$ gate inputs, and let $W$ be a set of tuples of constant faults on gate inputs of the circuit $S$. The *diagnosis problem for the circuit* $S$ *relative to the faults from* $W$: for a given circuit $S' \in \mathrm{Circ}(S, W)$ it is required to recognize the function realized by the circuit $S'$. To solve this problem we will use decision trees in which the computation of the value of each attribute consists in observation of output of the circuit $S'$ at the inputs of which a tuple from the set $\{0,1\}^n$ is given.

Define the diagnosis problem for the circuit $S$ relative to the faults from the set $W$ as a problem over corresponding information system. With each $\bar{\delta} \in \{0,1\}^n$ we associate the function $\bar{\delta} : \mathrm{Circ}(S, W) \to \{0,1\}$ such that $\bar{\delta}(S') = f_{S'}(\bar{\delta})$ for any $S' \in \mathrm{Circ}(S, W)$. Let us consider an information system $U(S, W) = (\mathrm{Circ}(S, W), \{0,1\}, \{0,1\}^n)$ and a problem $z_{S,W} = (\nu, \bar{\delta}_1, \ldots, \bar{\delta}_{2^n})$ over $U(S, W)$ where $\{\bar{\delta}_1, \ldots, \bar{\delta}_{2^n}\} = \{0,1\}^n$ and $\nu(\bar{\sigma}_1) \neq \nu(\bar{\sigma}_2)$ for any $\bar{\sigma}_1, \bar{\sigma}_2 \in \{0,1\}^{2^n}$ such that $\bar{\sigma}_1 \neq \bar{\sigma}_2$. The problem $z_{S,W}$ is a formalization of the notion of the diagnosis problem for the circuit $S$ relative to the faults from the set $W$.

The mapping $\nu$ from $z_{S,W}$ enumerates all Boolean functions of $n$ variables. The solution of the problem $z_{S,W}$ for a circuit $S' \in \mathrm{Circ}(S, W)$ is the number of function $f_{S'}$ realizing by the circuit $S'$. In some cases, it will be convenient for us instead of the number of the function $f_{S'}$ use a formula which realizes a function equal to $f_{S'}$.

Later, we will often consider the set $\{0,1,2\}^m$ of all possible tuples of constant faults on inputs of gates of the circuit $S$. Denote $U(S) = U(S, \{0,1,2\}^m)$, $z_S = z_{S,\{0,1,2\}^m}$ and $h(S) = h^g_{U(S)}(z_S)$. Evidently,

$h^g_{U(S)}(z_S) = h^l_{U(S)}(z_S)$. It is clear that $h(S)$ is the minimum depth of a decision tree over $U(S)$ solving the diagnostic problem for the circuit $S$ relative to the faults from the set $\{0, 1, 2\}^m$. If $S$ is a degenerate circuit then $h(S) = 0$.

### *10.2.2 Complexity of Decision Trees for Diagnosis of Faults*

In this subsection, the complexity of decision trees for diagnosis of arbitrary and specially constructed circuits is considered.

#### Arbitrary Circuits

The first direction of investigation is the study of the complexity of fault diagnosis algorithms (decision trees) for arbitrary circuits in the basis $B$. Let us consider for this purpose the function $h^{(1)}_B$ which characterizes the worst-case dependency of $h(S)$ on $\#(S)$ on the set $\mathrm{Circ}(B)$ of circuits. The function $h^{(1)}_B$ is defined in the following way:

$$h^{(1)}_B(n) = \max\{h(S) : S \in \mathrm{Circ}(B), \#(S) \le n\} .$$

The basis $B$ will be called *primitive* if at least one of the following conditions holds:

a) every function from $B$ is either a disjunction $x_1 \vee \ldots \vee x_n$ or a constant;

b) every function from $B$ is either a conjunction $x_1 \wedge \ldots \wedge x_n$ or a constant;

c) every function from $B$ is either a linear function $x_1 + \ldots + x_n + c \pmod 2$, $c \in \{0, 1\}$, or a constant.

**Theorem 10.8.** *For any basis $B$ the following statements hold:*
*a) if $B$ is a primitive basis then $h^{(1)}_B(n) = O(n)$;*
*b) if $B$ is a non-primitive basis then $\log_2 h^{(1)}_B(n) = \Omega(n^{1/2})$.*

The first part of the theorem statement is the most interesting for us. We now describe how it is possible to obtain the bound $h^{(1)}_B(n) = O(n)$ in the case when $B$ contains only linear functions and constants. Let $n \in \omega\setminus\{0\}$ and let $S$ be a circuit from $\mathrm{Circ}(B)$ with $\#(S) \le n$. Assume that $S$ is a nondegenerate circuit, and $S$ has exactly $r$ inputs labeled with variables $x_1, \ldots, x_r$ respectively. Denote by $m$ the number of gate inputs in the circuit $S$. Let exactly $t$ inputs of the circuit $S$ be linked by edges to gates, and let these inputs be labeled with variables $x_{i_1}, \ldots, x_{i_t}$ (possibly, $t = 0$). One can show that any circuit $S'$ from $\mathrm{Circ}(S, \{0, 1, 2\}^m)$ realizes a function of the kind

$(d_1 \wedge x_1) + \ldots + (d_r \wedge x_r) + d_0 (\text{mod } 2)$, where $d_j \in \{0,1\}$, $0 \le j \le r$. It is clear that $d_j = 0$ for any $j \in \{1, \ldots, r\} \setminus \{i_1, \ldots, i_t\}$.

Let us describe the work of a decision tree solving the problem $z_S$ which is the diagnosis problem for the circuit $S$ relative to the faults from the set $\{0,1,2\}^m$. Let $S' \in \text{Circ}(S, \{0,1,2\}^m)$. Give on the inputs of the circuit $S'$ the tuple consisting of zeros. We obtain at the output of the circuit $S'$ the value $d_0$. For each $j \in \{1, \ldots, t\}$, give some tuple on inputs of the circuit $S'$. Let $j \in \{1, \ldots, t\}$. Give the unity at the input of the circuit $S'$ labeled with the variable $x_{i_j}$, and give zeros at the other inputs of the circuit. We obtain value $d_{i_j} + d_0 (\text{mod } 2)$ at the output of the circuit. Thus, after the giving at the inputs of the circuit $S'$ of the considered $t + 1$ tuples, the coefficients $d_1, \ldots, d_r, d_0$ of the formula $(d_1 \wedge x_1) + \ldots + (d_r \wedge x_r) + d_0 (\text{mod } 2)$ will be recognized. Hence the considered decision tree solves the problem $z_S$, and the depth of this decision tree is at most $t + 1$. Therefore $h(S) \le t + 1$. Denote by $p$ the maximum number of variables in functions from $B$. It is clear that $t \le pn$. Set $c_1 = p + 1$. Then $h(S) \le c_1 n$. If $S$ is a degenerate circuit then $h(S) = 0 < c_1 n$. Taking into account that $S$ is an arbitrary circuit in the basis $B$ with $\#(S) \le n$ we obtain $h_B^{(1)}(n) \le c_1 n$. Therefore $h_B^{(1)}(n) = O(n)$.

The cases when $B$ contains only conjunctions and constants, or only disjunctions and constants can be considered in the same way.

## Specially Constructed Circuits

As opposed to the first one, the second direction of research explores complexity of diagnosis algorithms (decision trees) for circuits which are not arbitrary but chosen as the best from the point of view of solution of the diagnosis problem for the circuits, realizing the Boolean functions given as formulas over $B$. Let $\Phi(B)$ be the set of formulas over the basis $B$. For a formula $\varphi \in \Phi(B)$, we will denote by $\#(\varphi)$ the number of functional symbols in $\varphi$. Let $\varphi$ realize a function which does not belong to the set $\{0,1\}$. Set $h(\varphi) = \min h(S)$, where the minimum is taken over all possible combinatorial circuits $S$ (not necessarily in the basis $B$) which realize the same function as the formula $\varphi$. If $\varphi$ realizes a function from the set $\{0,1\}$ then $h(\varphi) = 0$. We will study the behavior of a function $h_B^{(2)}$ which characterizes the worst-case dependency of $h(\varphi)$ on $\#(\varphi)$ on the set of formulas over $B$ and is defined as follows:

$$h_B^{(2)}(n) = \max\{h(\varphi) : \varphi \in \Phi(B), \#(\varphi) \le n\} .$$

**Theorem 10.9.** *For an arbitrary basis $B$ the following statements hold:*

*a) if $B$ is a primitive basis then $h_B^{(2)}(n) = O(n)$;*

*b) if $B$ is a non-primitive basis then the equality $\log_2 h_B^{(2)}(n) = \Omega(n^c)$ holds for certain positive constant $c$ which depends only on $B$.*

### *10.2.3 Complexity of Construction of Decision Trees for Diagnosis*

The third direction of research is to study the complexity of algorithms for construction of decision trees for diagnosis problem.

A basis $B$ will be called *degenerate* if $B \subseteq \{0,1\}$, and *nondegenerate* otherwise. Let $B$ be a nondegenerate basis. Define an algorithmic problem $\mathrm{Con}(B)$.

The *problem* $\mathrm{Con}(B)$: for a given circuit $S$ from $\mathrm{Circ}(B)$ and a given set $W$ of tuples of constant faults on inputs of gates of the circuit $S$ it is required to construct a decision tree which solves the diagnosis problem for the circuit $S$ relative to the faults from $W$.

Note that there exists a decision tree which solves the diagnosis problem for the circuit $S$ relative to the faults from $W$ and the number of nodes in which is at most $2|W| - 1$.

**Theorem 10.10.** *Let $B$ be a nondegenerate basis. Then the following statements hold:*

*a) if $B$ is a primitive basis then there exists an algorithm which solves the problem* $\mathrm{Con}(B)$ *with polynomial time complexity;*

*b) if $B$ is a non-primitive basis then the problem* $\mathrm{Con}(B)$ *is NP-hard.*

### *10.2.4 Diagnosis of Iteration-Free Circuits*

From the point of view of the solution of the diagnosis problem for arbitrary tuples of constant faults on inputs of gates of arbitrary circuits, only primitive bases seem to be admissible. The extension of the set of such bases is possible by the substantial restriction on the class of the circuits under consideration. The fourth direction of research is the study of the complexity of fault diagnosis algorithms (decision trees) for iteration-free circuits.

Let $B$ be a basis. A circuit in the basis $B$ is called *iteration-free* if each node (input or gate) of it has at most one issuing edge. Let us denote by $\mathrm{Circ}^1(B)$ the set of iteration-free circuits in the basis $B$ with only one output. Let us consider the function $h_B^{(3)}$ which characterizes the worst-case dependency of $h(S)$ on $\#(S)$ for circuits from $\mathrm{Circ}^1(B)$ and is defined in the following way:

$$h_B^{(3)}(n) = \max\{h(S) : S \in \mathrm{Circ}^1(B), \#(S) \leq n\} .$$

Let us call a Boolean function $f(x_1, \ldots, x_n)$ *quasimonotone* if there exist numbers $\sigma_1, \ldots, \sigma_n \in \{0,1\}$ and a monotone Boolean function $g(x_1, \ldots, x_n)$ such that $f(x_1, \ldots, x_n) = g(x_1^{\sigma_1}, \ldots, x_n^{\sigma_n})$ where $x^\sigma = x$ if $\sigma = 1$, and $x^\sigma = \neg x$ if $\sigma = 0$.

The basis $B$ will be called *quasiprimitive* if at least one of the following conditions is true:

a) all functions from $B$ are linear functions or constants;
b) all functions from $B$ are quasimonotone functions.

The class of the quasiprimitive bases is rather large: for any basis $B_1$ there exists a quasiprimitive basis $B_2$ such that $\mathcal{F}(B_1) = \mathcal{F}(B_2)$, i.e., the set of Boolean functions realized by circuits in the basis $B_1$ coincides with the set of Boolean functions realized by circuits in the basis $B_2$.

**Theorem 10.11.** *Let $B$ be a basis. Then the following statements hold:*

*a) if $B$ is a quasiprimitive basis then $h_B^{(3)}(n) = O(n)$;*
*b) if $B$ is not a quasiprimitive basis then $\log_2 h_B^{(3)}(n) = \Omega(n)$.*

The first part of the theorem statement is the most interesting for us. The proof of this part is based on an efficient algorithm for diagnosis of iteration-free circuits in a quasiprimitive basis. Unfortunately, the description of this algorithm and the proof of its correctness are too long. However, we can illustrate the idea of algorithm. To this end, we consider another more simple problem of diagnosis [66].

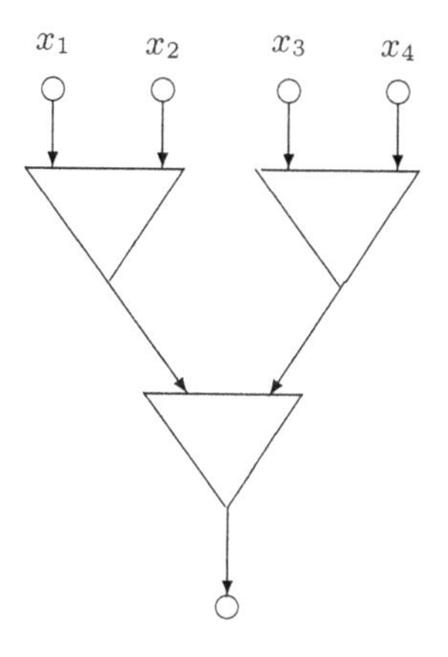


**Fig. 10.7**

Let we have an iteration-free circuit $S$ with one output in the basis $B = \{x \vee y, x \wedge y\}$. We know the "topology" of $S$ (corresponding directed acyclic graph) and variables attached to the inputs of $S$, but we do not know functions attached to gates (see, for example, a circuit $S_0$ depicted in Fig. 10.7). We should recognize functions attached to gates. To this end, we can give binary tuples at the inputs of the circuit and observe the output of the circuit. Note that if we give zeros on inputs of $S$, then at the output of $S$ we will have 0. If we give units at the inputs of $S$, then at the output of $S$ we will have 1.

Let $g_0$ be the gate of $S$ to which the output of $S$ is connected. Then there are two edges entering the gate $g_0$. These edges can be considered as outputs of two subcircuits of $S$—circuits $S_1$ and $S_2$. Let us give zeros at the inputs of $S_1$ and units at the inputs of $S_2$. If at the output of $S$ we have 0, then the function $\wedge$ is attached to the gate $g_0$. If at the output of $S$ we have 1, then the function $\vee$ is attached to the gate $g_0$.

Let the function $\wedge$ be attached to the gate $g_0$. We give units at the inputs of $S_1$. After that we can diagnose the subcircuit $S_2$: at the output of $S$ we will have the same value as at the output of $S_2$. The same situation is with the diagnosis of subcircuit $S_1$.

Let the function $\vee$ be attached to the gate $g_0$. We give zeros at the inputs of $S_1$. After that we can diagnose the subcircuit $S_2$: at the output of $S$ we will have the same value as at the output of $S_2$. The same situation is with the diagnosis of subcircuit $S_1$.

We see now that for the recognition of function attached to one gate we need to give at the inputs of $S$ one binary tuple and observe the output of $S$. So we can construct a decision tree for solving of the considered problem which depth is equal to $\#(S)$—the number of gates in $S$.

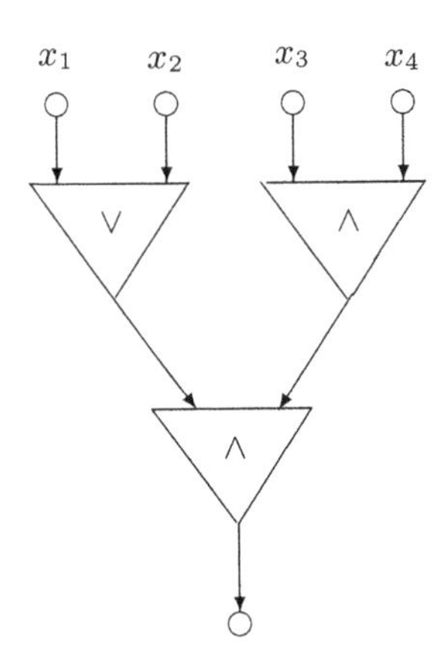


**Fig. 10.8**

*Example 10.12.* We now consider the circuit $S_0$ depicted in Fig. 10.7. Let us give at the inputs $x_1, x_2, x_3, x_4$ of $S_0$ the tuple $(0, 0, 1, 1)$, and let at the output we have 0. Then the function $\wedge$ is attached to the bottom gate of $S_0$. We now give the tuple $(0, 1, 1, 1)$ at the inputs of $S_0$, and let at the output we have 1. Then the the function $\vee$ is attached to the top left gate of $S_0$. Let us give the tuple $(1, 1, 0, 1)$ at the inputs of $S_0$, and let at the output we have 0. Then the function $\wedge$ is attached to the top right gate of $S_0$. As a result we obtain the circuit depicted in Fig. 10.8.

### *10.2.5 Approach to Circuit Construction and Diagnosis*

The fifth direction of research deals with the approach to the circuit construction and to the effective diagnosis of faults based on the results obtained for the iteration-free circuits. Two Boolean functions will be called *equal* if one of them can be obtained from the other by operations of insertion and deletion of unessential variables.

Using results from [91, 85] one can show that for each basis $B_1$ there exists a quasiprimitive basis $B_2$ with the following properties:

a) $\mathcal{F}(B_1) = \mathcal{F}(B_2)$, i.e., the set of functions realized by circuits in the basis $B_2$ coincides with the set of functions realized by circuits in the basis $B_1$;

b) there exists a polynomial $p$ such that for any formula $\varphi_1$ over $B_1$ there exists a formula $\varphi_2$ over $B_2$ which realizes the function equal to that realized by $\varphi_1$, and such that $\#(\varphi_2) \le p(\#(\varphi_1))$.

The considered approach to the circuit construction and fault diagnosis consists in the following. Let $\varphi_1$ be a formula over $B_1$ realizing certain function $f$, $f \notin \{0, 1\}$, and let us construct the formula $\varphi_2$ over $B_2$ realizing the function equal to $f$ and satisfying the inequality $\#(\varphi_2) \le p(\#(\varphi_1))$. Next a circuit $S$ in the basis $B_2$ is constructed (according to the formula $\varphi_2$) realizing the function $f$, satisfying the equality $\#(S) = \#(\varphi_2)$ and the condition that from each gate of the circuit $S$ at most one edge issues. In addition to the usual work mode of the circuit $S$ there exists the diagnostic mode in which the inputs of the circuit $S$ are "split" so that it becomes the iteration-free circuit $\tilde{S}$. From Theorem 10.11 it follows that the inequalities $h(\tilde{S}) \le c\#(S) \le cp(\#(\varphi_1))$, where $c$ is a constant depending only on the basis $B_2$, hold for the circuit $\tilde{S}$.

## 10.3 Conclusions

The chapter is devoted to the consideration of applications of theory of decision trees and decision rules to the problem of regular language word recognition and to the problem of diagnosis of constant faults in combinatorial circuits.

Proofs of the considered results are too complicated to be reproduced in this book. It should be noted that the most part of proofs (almost all can be found in [53]) is based on the bounds on complexity of decision trees and decision rule systems considered in Chap. 3.

Similar results for languages generated by some types of linear grammars and context-free grammars were obtained in [18, 28, 29].

We should mention three series of publications which are most similar to the results for diagnosis problem considered in this chapter. From the results obtained in [21, 27] the bound $h_B^{(3)}(n) = O(n)$ can be derived immediately for arbitrary basis $B$ with the following property: each function from $B$ is realized by some iteration-free circuit in the basis $\{x \wedge y, x \vee y, \neg x\}$. In [74, 75, 76, 77], for circuits in an arbitrary finite basis and faults of different types (not only the constant) the dependence is investigated of the minimum depth of a decision tree, which diagnoses circuit faults, on total number of inputs and gates in the circuit. In [56, 64, 65, 78], effective methods for diagnosis of faults of different types are considered.

# Final Remarks

This book is oriented to the use of decision trees and decision rule systems not only as predictors but also as algorithms and ways for knowledge representation.

The main aims of the book are (i) to describe a set of tools that allow us to work with exact and approximate decision trees, decision rule systems and reducts (tests) for usual decision tables and decision tables with many-valued decisions, and (ii) to give a number of examples of the use of these tools in such areas of applications as supervised learning, discrete optimization, analysis of acyclic programs, pattern recognition and fault diagnosis.

Usually, we have no possibility to give proofs for statements connected with applications—proofs are too long and complicated. However, when it is possible, we add comments connected with the use of tools from the first part of the book. In contrast to applications, almost all statements relating to tools are given with simple and short proofs.

In the book, we concentrate on the consideration of time complexity in the worst case of decision trees (depth) and decision rule systems (maximum length of a rule in the system). In the last case, we assume that we can work with rules in parallel. The problems of minimization of average time complexity (average depth) or space complexity (number of nodes) of decision trees are essentially more complicated. However, we can generalize some results considered in the book to these cases (in particular, dynamic programming approach to optimization of decision trees). The problems of optimization of average time complexity of decision rule systems (average length of rules) and space complexity (number of rules or total length of rules) are even more complicated.

We consider not only decision tables and finite information systems but also study infinite information systems in the frameworks of both local and global approaches. The global approach is essentially more complicated than the local one: we need to choose appropriate attributes from an infinite set of attributes. However, as a result, we often can find decision trees and decision rule systems with relatively small time complexity in the worst case.

# References


1. Aha, D.W. (ed.): Lazy Learning. Kluwer Academic Publishers, Dordrecht (1997)
2. Alkhalid, A., Chikalov, I., Moshkov, M.: On algorithm for building of optimal $\alpha$-decision trees. In: Szczuka, M., Kryszkiewicz, M., Ramanna, S., Jensen, R., Hu, Q. (eds.) RSCTC 2010. LNCS, vol. 6086, pp. 438–445. Springer, Heidelberg (2010)
3. Bazan, J.G.: Discovery of decision rules by matching new objects against data tables. In: Polkowski, L., Skowron, A. (eds.) RSCTC 1998. LNCS (LNAI), vol. 1424, pp. 521–528. Springer, Heidelberg (1998)
4. Bazan, J.G.: A comparison of dynamic and non-dynamic rough set methods for extracting laws from decision table. In: Polkowski, L., Skowron, A. (eds.) Rough Sets in Knowledge Discovery 1. Methodology and Applications. Studies in Fuzziness and Soft Computing, vol. 18, pp. 321–365. Phisica-Verlag, Heidelberg (1998)
5. Bazan, J.G.: Methods of approximate reasoning for synthesis of decision algorithms. Ph.D. Thesis, Warsaw University, Warsaw (1998) (in Polish)
6. Boros, E., Hammer, P.L., Ibaraki, T., Kogan, A.: Logical analysis of numerical data. Math. Programming 79, 163–190 (1997)
7. Boros, E., Hammer, P.L., Ibarki, T., Kogan, A., Mayoraz, E., Muchnik, I.: An implementation of logical analysis of data. IEEE Transactions of Knowledge and Data Engineering 12, 292–306 (2000)
8. Breiman, L., Friedman, J.H., Olshen, R.A., Stone, C.J.: Classification and Regression Trees. Chapman and Hall, New York (1984)
9. Cheriyan, J., Ravi, R.: Lecture notes on approximation algorithms for network problems (1998), `http://www.math.uwaterloo.ca/~jcheriya/lecnotes.html`
10. Chikalov, I.: Algorithm for constructing of decision trees with minimal average depth. In: Proc. Eighth Int'l Conference on Information Processing and Management of Uncertainty in Knowledge-based Systems, Madrid, Spain, vol. 1, pp. 376–379 (2000)
11. Chikalov, I.V.: Algorithm for constructing of decision trees with minimal number of nodes. In: Ziarko, W.P., Yao, Y. (eds.) RSCTC 2000. LNCS (LNAI), vol. 2005, pp. 139–143. Springer, Heidelberg (2001)

12. Chikalov, I.V., Moshkov, M.J., Zelentsova, M.S.: On optimization of decision trees. In: Peters, J.F., Skowron, A. (eds.) Transactions on Rough Sets IV. LNCS, vol. 3700, pp. 18–36. Springer, Heidelberg (2005)
13. Chikalov, I., Moshkov, M., Zielosko, B.: Upper bounds on minimum cardinality of reducts and depth of decision trees for decision tables with many-valued decisions. In: Proc. Concurrency, Specification and Programming, Helenenau, Germany, pp. 97–103 (2010)
14. Chikalov, I., Moshkov, M., Zielosko, B.: Upper bounds on minimum cardinality of exact and approximate reducts. In: Szczuka, M., Kryszkiewicz, M., Ramanna, S., Jensen, R., Hu, Q. (eds.) RSCTC 2010. LNCS, vol. 6086, pp. 412–417. Springer, Heidelberg (2010)
15. Chlebus, B.S., Nguyen, S.H.: On finding optimal discretizations for two attributes. In: Polkowski, L., Skowron, A. (eds.) RSCTC 1998. LNCS (LNAI), vol. 1424, pp. 537–544. Springer, Heidelberg (1998)
16. Cover, T.M., Hart, P.E.: Nearest neighbor pattern classification. IEEE Transactions on Information Theory 13, 21–27 (1967)
17. Crama, Y., Hammer, P.L., Ibaraki, T.: Cause-effect relationships and partially defined Boolean functions. Ann. Oper. Res. 16, 299–326 (1988)
18. Dudina, J.V., Knyazev, A.N.: On complexity of recognition of words from languages generated by context-free grammars with one nonterminal symbol. Vestnik of Lobachevsky State University of Nizhni Novgorod 2, 214–223 (1998)
19. Feige, U.: A threshold of $\ln n$ for approximating set cover (Preliminary version). In: Proc. 28th Annual ACM Symposium on the Theory of Computing, pp. 314–318 (1996)
20. Friedman, J.H., Kohavi, R., Yun, Y.: Lazy decision trees. In: Proc. 13th National Conference on Artificial Intelligence, pp. 717–724. AAAI Press, Menlo Park (1996)
21. Goldman, R.S., Chipulis, V.P.: Diagnosis of iteration-free combinatorial circuits. In: Zhuravlev, J.I. (ed.) Discrete Analysis, vol. 14, pp. 3–15. Nauka Publishers, Novosibirsk (1969) (in Russian)
22. Goldman, S., Kearns, M.: On the complexity of teaching. In: Proc. 1th Annual Workshop on Computational Learning Theory, pp. 303–314 (1991)
23. Hastie, T., Tibshirani, R., Friedman, J.: The Elements of Statistical Learning: Data Mining, Inference, and Prediction. Springer, New York (2001)
24. Hegedüs, T.: Generalized teaching dimensions and the query complexity of learning. In: Proc. 8th Annual ACM Conference on Computational Learning Theory, pp. 108–117 (1995)
25. Hellerstein, L., Pillaipakkamnatt, K., Raghavan, V.V., Wilkins, D.: How many queries are needed to learn? J. ACM 43, 840–862 (1996)
26. Johnson, D.S.: Approximation algorithms for combinatorial problems. J. Comput. System Sci. 9, 256–278 (1974)
27. Karavai, M.F.: Diagnosis of tree-like circuits in arbitrary basis. Automation and Telemechanics 1, 173–181 (1973) (in Russian)
28. Knyazev, A.: On recognition of words from languages generated by linear grammars with one nonterminal symbol. In: Polkowski, L., Skowron, A. (eds.) RSCTC 1998. LNCS (LNAI), vol. 1424, pp. 111–114. Springer, Heidelberg (1998)

29. Knyazev, A.N.: On recognition of words from languages generated by context-free grammars with one nonterminal symbol. In: Proc. Eighth Int'l Conference on Information Processing and Management of Uncertainty in Knowledge-based Systems, Madrid, Spain, vol. 1, pp. 1945–1948 (2000)
30. Laskowski, M.C.: Vapnik-Chervonenkis classes of definable sets. J. London Math. Society 45, 377–384 (1992)
31. Littlestone, N.: Learning quickly when irrelevant attributes abound: a new linear threshold algorithm. Machine Learning 2, 285–318 (1988)
32. Markov, A.: Introduction into Coding Theory. Nauka Publishers, Moscow (1982)
33. Meyer auf der Heide, F.: A polynomial linear search algorithm for the $n$-dimensional knapsack problem. J. ACM 31, 668–676 (1984)
34. Meyer auf der Heide, F.: Fast algorithms for $n$-dimensional restrictions of hard problems. J. ACM 35, 740–747 (1988)
35. Moshkov, M.: About uniqueness of uncancellable tests for recognition problems with linear decision rules. In: Markov, A. (ed.) Combinatorial-Algebraic Methods in Applied Mathematics, pp. 97–109. Gorky University Publishers, Gorky (1981) (in Russian)
36. Moshkov, M.: On conditional tests. Academy of Sciences Doklady 265, 550–552 (1982) (in Russian); English translation: Sov. Phys. Dokl. 27, 528–530 (1982)
37. Moshkov, M.: Conditional tests. In: Yablonskii, S.V. (ed.) Problems of Cybernetics, vol. 40, pp. 131–170. Nauka Publishers, Moscow (1983) (in Russian)
38. Moshkov, M.: Elements of mathematical theory of tests (methodical indications). Gorky State University, Gorky (1986)
39. Moshkov, M.: Elements of mathematical theory of tests, part 2 (methodical development). Gorky State University, Gorky (1987)
40. Moshkov, M.: On relationship of depth of deterministic and nondeterministic acyclic programs in the basis $\{x + y, x - y, 1; \text{sign } x\}$. In: Mathematical Problems in Computation Theory, Banach Center Publications, vol. 21, pp. 523–529. Polish Scientific Publishers, Warsaw (1988)
41. Moshkov, M.: Decision Trees. Theory and Applications. Theory and Applications. Nizhny Novgorod University Publishers, Nizhny Novgorod (1994) (in Russian)
42. Moshkov, M.: Decision trees with quasilinear checks. Trudy IM SO RAN 27, 108–141 (1994) (in Russian)
43. Moshkov, M.: Unimprovable upper bounds on complexity of decision trees over information systems. Foundations of Computing and Decision Sciences 21, 219–231 (1996)
44. Moshkov, M.: On global Shannon functions of two-valued information systems. In: Proc. Fourth Int'l Workshop on Rough Sets, Fuzzy Sets and Machine Discovery, Tokyo, Japan, pp. 142–143 (1996)
45. Moshkov, M.: Lower bounds for the time complexity of deterministic conditional tests. Diskr. Mat. 8, 98–110 (1996) (in Russian)
46. Moshkov, M.: Complexity of deterministic and nondeterministic decision trees for regular language word recognition. In: Proc. Third Int'l Conference Developments in Language Theory, Thessaloniki, Greece, pp. 343–349 (1997)

47. Moshkov, M.: On time complexity of decision trees. In: Polkowski, L., Skowron, A. (eds.) Rough Sets in Knowledge Discovery 1. Methodology and Applications. Studies in Fuzziness and Soft Computing, vol. 18, pp. 160–191. Phisica-Verlag, Heidelberg (1998)
48. Moshkov, M.: Diagnosis of constant faults in circuits. In: Lupanov, O.B. (ed.) Mathematical Problems of Cybernetics, vol. 9, pp. 79–100. Nauka Publishers, Moscow (2000)
49. Moshkov, M.: Elements of Mathematical Theory of Tests with Applications to Problems of Discrete Optimization: Lectures. Nizhny Novgorod University Publishers, Nizhny Novgorod (2001) (in Russian)
50. Moshkov, M.: Classification of infinite information systems depending on complexity of decision trees and decision rule systems. Fundam. Inform. 54, 345–368 (2003)
51. Moshkov, M.J.: Greedy algorithm of decision tree construction for real data tables. In: Peters, G.F., Skowron, A., Grzymala-Busse, J.W., Kostek, B., Swiniarski, R.W., Szczuka, M.S. (eds.) Transactions on Rough Sets I. LNCS, vol. 3100, pp. 161–168. Springer, Heidelberg (2004)
52. Moshkov, M.J.: Greedy algorithm for decision tree construction in context of knowledge discovery problems. In: Tsumoto, S., Słowiński, R., Komorowski, J., Grzymała-Busse, J.W. (eds.) RSCTC 2004. LNCS (LNAI), vol. 3066, pp. 192–197. Springer, Heidelberg (2004)
53. Moshkov, M.J.: Time complexity of decision trees. In: Peters, J.F., Skowron, A. (eds.) Transactions on Rough Sets III. LNCS, vol. 3400, pp. 244–459. Springer, Heidelberg (2005)
54. Moshkov, M.: On the class of restricted linear information systems. Discrete Mathematics 307, 2837–2844 (2007)
55. Moshkov, M., Chikalov, I.: On algorithm for constructing of decision trees with minimal depth. Fundam. Inform. 41, 295–299 (2000)
56. Moshkov, M., Moshkova, A.: Optimal bases for some closed classes of Boolean functions. In: Proc. Fifth European Congress on Intelligent Techniques and Soft Computing, Aachen, Germany, pp. 1643–1647 (1997)
57. Moshkov, M.J., Piliszczuk, M., Zielosko, B.: On partial covers, reducts and decision rules with weights. In: Peters, J.F., Skowron, A., Düntsch, I., Grzymała-Busse, J.W., Orłowska, E., Polkowski, L. (eds.) Transactions on Rough Sets VI. LNCS, vol. 4374, pp. 211–246. Springer, Heidelberg (2007)
58. Moshkov, M., Piliszczuk, M., Zielosko, B.: On construction of partial reducts and irreducible partial decision rules. Fundam. Inform. 75, 357–374 (2007)
59. Moshkov, M., Piliszczuk, M., Zielosko, B.: Partial Covers, Reducts and Decision Rules in Rough Sets: Theory and Applications. SCI, vol. 145. Springer, Heidelberg (2008)
60. Moshkov, M., Piliszczuk, M., Zielosko, B.: On partial covers, reducts and decision rules. In: Peters, J.F., Skowron, A. (eds.) Transactions on Rough Sets VIII. LNCS, vol. 5084, pp. 251–288. Springer, Heidelberg (2008)
61. Moshkov, M., Piliszczuk, M., Zielosko, B.: Universal problem of attribute reduction. In: Peters, J.F., Skowron, A., Rybiński, H. (eds.) Transactions on Rough Sets IX. LNCS, vol. 5390, pp. 187–199. Springer, Heidelberg (2008)
62. Moshkov, M., Piliszczuk, M., Zielosko, B.: Greedy algorithm for construction of partial association rules. Fundam. Inform. 92, 259–277 (2009)
63. Moshkov, M., Piliszczuk, M., Zielosko, B.: Greedy algorithms with weights for construction of partial association rules. Fundam. Inform. 94, 101–120 (2009)

64. Moshkova, A.M.: On diagnosis of "retaining" faults in circuits. In: Polkowski, L., Skowron, A. (eds.) RSCTC 1998. LNCS (LNAI), vol. 1424, pp. 513–516. Springer, Heidelberg (1998)
65. Moshkova, A.M.: On time complexity of "retaining" fault diagnosis in circuits. In: Proc. Eighth Int'l Conference on Information Processing and Management of Uncertainty in Knowledge-based Systems, Madrid, Spain, vol. 1, pp. 372–375 (2000)
66. Moshkova, A., Moshkov, M.: Unpublished manuscript
67. Nguyen, H.S., Ślęzak, D.: Approximate reducts and association rules—correspondence and complexity results. In: Zhong, N., Skowron, A., Ohsuga, S. (eds.) RSFDGrC 1999. LNCS (LNAI), vol. 1711, pp. 137–145. Springer, Heidelberg (1999)
68. Nigmatullin, R.G.: Method of steepest descent in problems on cover. In: Memoirs of Symposium Problems of Precision and Efficiency of Computing Algorithms, Kiev, USSR, vol. 5, pp. 116–126 (1969) (in Russian)
69. Nowak, A., Zielosko, B.: Inference processes on clustered partial decision rules. In: Kłopotek, M.A., Przepiórkowski, A., Wierzchoń, S.T. (eds.) Recent Advances in Intelligent Information Systems, pp. 579–588. Academic Publishing House EXIT, Warsaw (2009)
70. Pawlak, Z.: Rough Sets—Theoretical Aspects of Reasoning about Data. Kluwer Academic Publishers, Dordrecht (1991)
71. Quinlan, J.R.: C4.5: Programs for Machine Learning. Morgan Kaufmann, San Mateo (1993)
72. Rissanen, J.: Modeling by shortest data description. Automatica 14, 465–471 (1978)
73. Rough Set Exploration System (RSES), `http://logic.mimuw.edu.pl/~rses`
74. Shevtchenko, V.: On depth of conditional tests for diagnosis of "negation" type faults in circuits. Siberian Journal on Operations Research 1, 63–74 (1994) (in Russian)
75. Shevtchenko, V.: On the depth of decision trees for diagnosing faults in circuits. In: Lin, T.Y., Wildberger, A.M. (eds.) Soft Computing, pp. 200–203. Society for Computer Simulation, San Diego, California (1995)
76. Shevtchenko, V.: On the depth of decision trees for control faults in circuits. In: Proc. Fourth Int'l Workshop on Rough Sets, Fuzzy Sets and Machine Discovery, Tokyo, Japan, pp. 328–330 (1996)
77. Shevtchenko, V.: On the depth of decision trees for diagnosing of nonelementary faults in circuits. In: Polkowski, L., Skowron, A. (eds.) RSCTC 1998. LNCS (LNAI), vol. 1424, pp. 517–520. Springer, Heidelberg (1998)
78. Shevtchenko, V., Moshkov, M., Moshkova, A.: Effective methods for diagnosis of faults in circuits. In: Proc. 11th Interstates Workshop Design and Complexity of Control Systems, Nizhny Novgorod, Russia, vol. 2, pp. 228–238 (2001) (in Russian)
79. Skowron, A.: Rough sets in KDD. In: Shi, Z., Faltings, B., Musen, M. (eds.) Proc. 16th IFIP World Computer Congress, pp. 1–14. Publishing House of Electronic Industry, Beijing (2000)
80. Skowron, A., Rauszer, C.: The discernibility matrices and functions in information systems. In: Slowinski, R. (ed.) Intelligent Decision Support. Handbook of Applications and Advances of the Rough Set Theory, Kluwer Academic Publishers, Dordrecht (1992)

81. Slavík, P.: A tight analysis of the greedy algorithm for set cover. In: Proc. 28th Annual ACM symposium on the theory of computing, pp. 435–441. ACM Press, New York (1996)
82. Slavík, P.: Approximation algorithms for set cover and related problems. Ph.D. Thesis, University of New York at Buffalo (1998)
83. Ślęzak, D.: Approximate entropy reducts. Fundam. Inform. 53, 365–390 (2002)
84. Soloviev, N.A.: Tests (Theory, Construction, Applications). Nauka Publishers, Novosibirsk (1978) (in Russian)
85. Ugolnikov, A.B.: On depth and polynomial equivalence of formulae for closed classes of binary logic. Mathematical Notes 42, 603–612 (1987) (in Russian)
86. Vapnik, V.N.: Statistical Learning Theory. Wiley, New York (1998)
87. Vapnik, V.N., Chervonenkis, A.Y.: On the uniform convergence of relative frequencies of events to their probabilities. Theory of Probability and its Applications 16, 264–280 (1971)
88. Wróblewski, J.: Ensembles of classifiers based on approximate reducts. Fundam. Inform. 47, 351–360 (2001)
89. Yablonskii, S.V.: Tests. In: Glushkov, V.M. (ed.) Encyklopaedia Kybernetiki, pp. 431–432. Main Editorial Staff of Ukrainian Soviet Encyklopaedia, Kiev (1975) (in Russian)
90. Yablonskii, S.V., Chegis, I.A.: On tests for electric circuits. Uspekhi Matematicheskikh Nauk 10, 182–184 (1955) (in Russian)
91. Yablonskii, S.V., Gavrilov, G.P., Kudriavtzev, V.B.: Functions of Algebra of Logic and Classes of Post. Nauka Publishers, Moscow (1966) (in Russian)
92. Zhuravlev, J.I.: On a class of partial Boolean functions. In: Zhuravlev, J.I. (ed.) Discretny Analis, vol. 2, pp. 23–27. IM SO AN USSR Publishers, Novosibirsk (1964) (in Russian)
93. Zielosko, B.: On partial decision rules. In: Proc. Concurrency, Specification and Programming, Ruciane-Nida, Poland, pp. 598–609 (2005)
94. Zielosko, B.: Greedy algorithm for construction of partial association rules. Studia Informatica 31, 225–236 (2010) (in Polish)
95. Zielosko, B., Piliszczuk, M.: Greedy algorithm for attribute reduction. Fundam. Inform. 85, 549–561 (2008)
96. Zielosko, B., Marszał-Paszek, B., Paszek, P.: Partial and nondeterministic decision rules in classification process. In: Kłopotek, M.A., Przepiórkowski, A., Wierzchoń, S.T. (eds.) Recent Advances in Intelligent Information Systems, pp. 629–638. Academic Publishing House EXIT, Warsaw (2009)
97. Zielosko, B., Moshkov, M., Chikalov, I.: Decision rule optimization on the basis of dynamic programming methods. Vestnik of Lobachevsky State University of Nizhni Novgorod 6, 195–200 (2010)

# Index